Wetland Creations, Restoration and Conversation
The State of Science

Wetland Creations, Restoration and Conversation
The State of Science

Edited by William J. Mitsch

The Ohio State University
Columbus Ohio
USA

This volume is reprinted from Ecology Engineering, volume 24, issue 4, 5 April 2005

ELSEVIER

Amsterdam • Boston • Heidelberg • London • New York • Oxford
Paris • San Diego • San Francisco • Singapore • Sydney • Tokyo

ELSEVIER B.V.
Radarweg 29
P.O. Box 211, 1000 AE
Amsterdam, The Netherlands

ELSEVIER Inc.
525 B Street, Suite 1900
San Diego, CA 92101-4495
USA

ELSEVIER Ltd
The Boulevard, Langford Lane
Kidlington, Oxford OX5 1GB
UK

ELSEVIER Ltd
84 Theobalds Road
London WC1X 8RR
UK

First edition 2005

Library of Congress Cataloging in Publication Data
A catalog record is available from the Library of Congress.

British Library Cataloguing in Publication Data
A catalogue record is available from the British Library.

ISBN: 0-444-52134-8

⊗ The paper used in this publication meets the requirements of ANSI/NISO Z39.48-1992 (Permanence of Paper).
Printed in The Netherlands.

ELSEVIER

Ecological Engineering 24 (2005) V

ECOLOGICAL ENGINEERING

www.elsevier.com/locate/ecoleng

Contents

Available online at www.sciencedirect.com

Ecological Engineering 24 (2005) 243–251

ECOLOGICAL ENGINEERING

www.elsevier.com/locate/ecoleng

Editorial

Wetland creation, restoration, and conservation: A Wetland Invitational at the Olentangy River Wetland Research Park

William J. Mitsch *

Olentangy River Wetland Research Park, The Ohio State University, 352 West Dodridge Street, Columbus, OH 43202, USA

1. Introduction

Wetlands are shallow to intermittently flooded ecosystems that are more commonly known by such terms as swamps, bogs, marshes, and sedge meadows. They are revered as important parts of the natural landscape because of their functions in cleaning and retaining water naturally, preventing floods, and providing a habitat and food source for a wide variety of plant and animal species. It is estimated that more than half of the original wetlands in the world have been lost to drainage projects and human development projects. The state where our Wetlands Invitational was held, Ohio, has lost about 90% of its original wetlands. Thus, both wetland protection and the ability to create and restore wetlands and their supporting aquatic environments are vital areas for continued study.

When we lose wetlands, we lose their ability to provide clean water, prevent floods, and enhance biological diversity. Many organizations are calling for restoration and creation of wetlands to clean up our streams, rivers, and lakes. The US National Academy of Sciences called for the restoration and creation of 4 million ha of wetlands in the United States by the year 2010. Creation and restoration of several million

hectares of wetlands have been recommended for the Mississippi River Basin to help prevent the dead zone, or hypoxia, in the Gulf of Mexico (Mitsch et al., 2001; Mississippi River Basin Task Force, 2001). The US Army Corps of Engineers oversees a regulatory program that results in thousands of hectares of wetlands being restored and created each year to replace wetlands that are lost to development. Futhermore, the largest wetland and riverine restorations in the world, at costs that will exceed US$ 20 billion, are underway or planned for the Everglades and Louisiana Delta. Yet a National Academy of Sciences panel (NRC, 2001) determined that much more research is needed before we can be assured that wetlands created to replace wetlands destroyed for development can be successful. In order to solve such problems, we need to know: (1) how wetlands work; (2) if we can create and restore them; and (3) the best approaches to creation and restoration of wetlands. The Olentangy River Wetland Research Park (ORWRP) is designed to be a long-term, large-scale wetland research facility on a major university campus.

This editorial celebrates the research, teaching, and completion of the master plan of the Olentangy River Wetland Research Park at The Ohio State University in Columbus, USA, and describes a series of 11 papers that are part of this special issue of *Ecological Engineering* that resulted from a scientific event enti-

* Tel.: +1 614 2929774; fax: +1 614 2929773.
 E-mail address: mitsch.1@osu.edu.

0925-8574/$ – see front matter © 2005 Published by Elsevier B.V.
doi:10.1016/j.ecoleng.2005.02.006

tled "Wetland Invitational" that brought some of the world's best wetland scientists to Ohio on May 15–16, 2003, to participate in that celebration.

2. The Olentangy River Wetland Research Park

The cause for the Wetland Invitational was the completion of the Olentangy River Wetland Research Park in Ohio. The ORWRP (Figs. 1 and 2) is located on the campus of The Ohio State University in Columbus, OH, USA. Through its 12 years of development and research activity, the ORWRP had as its goals:

1. Research on ecological processes in created and naturally occurring wetlands, particularly for improving water quality, mitigating flooding, and providing habitat.
2. Research on the proper design criteria for wetlands and on measuring success of these wetlands.
3. Graduate and undergraduate teaching in subjects related to water resources, rivers, and wetlands.
4. Continuing education and demonstrations on wetland topics for public agencies, private consultants, and the public.

This research and teaching site has been developed in three phases. Phase 1 of site development featured construction of two 1-ha deepwater marshes and a river water delivery system and was completed in 1994. Pumps were installed on the floodplain to bring water from the Olentangy River to the two kidney-shaped experimental wetlands and river water began to flow to the wetlands on March 4, 1994. River water is pumped continuously, day and night, into the two wetlands. It then flows by gravity back to the Olentangy River through a swale and constructed stream system. In May 1994, one wetland basin was planted with marsh vegetation typical of wetlands in the Midwest; the other remained as an unplanted control (see Mitsch et al., 1998, 2005a). Phase 2, establishing the infrastructure for research and education of the site, began in 1994 and was completed with the dedication of the Sandefur Wetland Pavilion in 1999. This phase included construction of plastic boardwalks in the two experimental wetlands, construction of a 2.5-ha floodplain wetland (called a billabong), and construction of a mesocosm compound for small-scale experimental research replicated in outdoor ponds. Planning for Phase 3, the construction of the Heffner Wetland Research and Education Building, began with the receipt of US$ 1.2 million from the

Fig. 1. Olentangy River Wetland Research Park at The Ohio State University. Features are identified on Fig. 2. North direction is to the left.

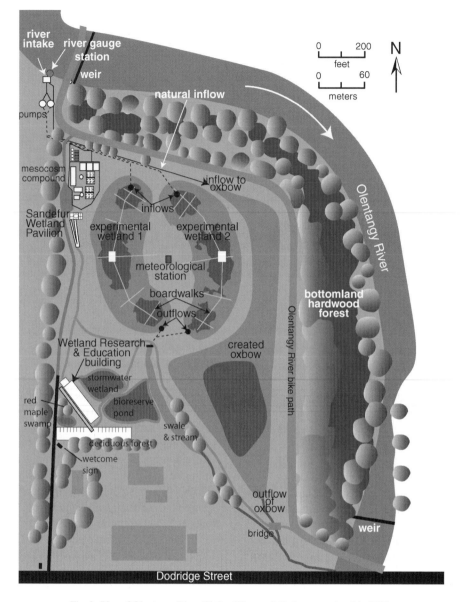

Fig. 2. Plan of Olentangy River Wetland Research Park as completed in 2003.

Ohio Board of Regents in 1999 in their Hayes Invest-ment Fund competition. The grants were the result of an effort of a consortium of five Ohio institutions—Ohio State, Wright State, Shawnee State, Youngstown State, and Kenyon College. Additional support for the build-ing was obtained through donations and pledges from many individuals and organizations. Building con-struction began in spring 2002 and researchers moved

into the building that was subsequently named the Heffner Wetland Research and Education Building (Fig. 3) by March 2003. A decision was made to have a scientific celebration for that completion in May 2003; this special issue is the result of that celebration.

The contributions of the ORWRP from 1991 through 2004 to The Ohio State University and to stu-dents, the public, and the world of wetland science have

Fig. 3. Heffner Wetland Research and Education Building, Olentangy River Wetland Research Park, The Ohio State University. The Heffner building was the site of the Wetland Invitational held on May 15–16, 2003.

been significant:

- completion of 44 undergraduate and graduate student theses and dissertations on wetlands and related subjects, including five theses from European institutions;
- publication of 113 papers listed in the ORWRP reprint series (listed at http://swamp.osu.edu);
- completion of 12 comprehensive annual reports summarizing all research accomplished at the OR-WRP;
- leadership of almost formal wetland tours and presentations for the public to an estimated 18,000 individuals including K-12 students, university students, garden clubs, campus visitors, and Federal, state, and local public officials;
- provision of a convenient set of campus ecosystems in support of an estimated 160 Ohio State University classes and activities from seven university colleges;
- education of almost 300 agency personnel and consultants in 20 wetland short courses since 1996;
- development of the fields of wetland science and ecological engineering to the point where they have led to a significant improvement in Ohio's, the nation's, and the world's environment.

3. The Wetland Invitational

Twelve of the world's best wetland scientists were invited to the ORWRP to present papers at our Wetland Invitational related to wetland restoration and preservation throughout the world (Fig. 4; Table 1). The distinguished group of scientists and about 150 participants in the audience were welcomed to Ohio State University by OSU President Karen Holbrook (Fig. 5). All 12 speakers were awarded plaques that designated their talks as "Moonlight on the Marsh Distinguished Lectures," and a permanent plaque in the Heffner Wetland building lobby includes a list of these talks. Discussions on wetland creation, restoration, and conservation at the Invitational were lively.

4. This special issue

This special issue contains 11 papers that, from the most part, were presented by the participants at the Wetland Invitational in Columbus. They are divided, by subject matter into four general categories:

1. Restoration of a large river basin and delta.
2. Long-term wetland restoration.

Fig. 4. Invited speakers at the Wetlands Invitational; left to right, Curt Richardson, Duke University; Rob Brooks, Penn State University; Mark Brown, University of Florida; Bill Mitsch, Ohio State University; John Day, Louisiana State University; Dennis Whigham, Smithsonian Environmental Research Center; Ed Garbisch, Environmental Concern Inc.; Siobhan Fennessy, Kenyon College; Don Hey, The Wetlands Initiative; Wolfgang Junk, Max Planck Institute; Jim Amon, Wright State University; Tom Crisman, University of Florida.

Fig. 5. Ohio State University President Karen Holbrook welcomes participants to the Wetland Invitational in Columbus, OH, USA, on May 15–16, 2003.

Table 1
Guest lectures and panel discussions at Wetlands Invitational at Olentangy River Wetland Research Park, The Ohio State University, Columbus, OH, USA, May 15–16, 2003

Thursday May 15

Guest lectures

John W. Day Jr., Distinguished Professor, Coastal Ecology Institute, Louisiana State University "Restoration of the Mississippi Delta"

William J. Mitsch, Distinguished Professor of Natural Resources and Environmental Science and Director, Olentangy River Wetland Research Park, The Ohio State University "Restoring the Mississippi River Basin with 10 million ha of wet lands"

Dennis Whigham, Smithsonian Environmental Research Center, Maryland "Slowing the rate of degradation in the Chesapeake Bay through wetland restoration"

Donald Hey, Senior Vice-President, The Wetlands Initiative, Chicago "Financing wetland restoration"

M. Siobhan Fennessy, Associate Professor, Kenyon College "Planning wetland restoration at the watershed scale: lessons from the Cuyahoga River Basin, OH"

Edgar W. Garbisch, President, Environmental Concern Inc., St. Michaels, MD "Coastal wetland creation at Hambleton Island, Chesapeake Bay, Maryland in 1971, and years later"

Panel discussion—research needs for integrating coastal and watershed restoration

Champagne celebration

Celebration of completion of the home for the "Ohio Center for Wetland and River Restoration" at the Olentangy River Wetland Research Park

Friday May 16

Guest lectures

Curtis Richardson, Professor and Director Duke Wetland Center, Nicholas School of Environment and Earth Sciences, Duke University "Successful Everglades restoration is not a river of grass"

Robert P. Brooks, Professor of Wildlife and Wetlands and Director of the Penn State Cooperative Wetlands Center. Penn State University "Are we purveyors of wetland homogeneity?: a model of degradation and restoration to improve mitigation performance"

James Amon, Associate Professor of Biology, Wright State University "For peat's sake—restore the wetlands"

Mark T. Brown, Associate Professor of Environmental Engineering Sciences, Howard T. Odum Center for Wetlands, University of Florida "Landscape restoration: Insights and design principles gained from 25 years of co-evolution of science, industry, and regulation related to Florida's phosphate mining"

Thomas Crisman, Director, Howard T. Odum Center for Wetlands and Professor of Environmental Engineering Sciences, University of Florida "How little water does a wetland need to function? The reality of transboundary conflicts and water scarcity in the Mediterranean Basin, Middle East, and Africa"

Wolfgang Junk, Max Planck Institute, Plon, Germany, and Manaus, Brazil "Amazonian wetlands: Classification, distribution, sustainable management, and threats"

Panel discussion—conserving and restoring the world's wetlands—research needs

Lunch event (sponsored by OSU Environmental Science Graduate Program)

3. Creation of wetlands for mitigation of wetland loss.
4. Conservation and restoration of the world's wetlands.

4.1. Restoration of a large river basin and delta

There are two major wetland restoration efforts being discussed in the United States regarding the Mississippi River Basin, a 3 million km^2 river basin that comprised 40% of the lower 48 states in the United States. In the delta of the Mississippi River in Louisiana, wetlands are being lost at rates of 60–100 km^2/year with 4800 km^2 of wetlands lost since the 1930s. The principal, but not only, cause is the flood-control levees along the Mississippi River that prevent the river from spreading floodwaters over the delta but instead shunt them out into the abyss of the Gulf of Mexico. Day et al. (2005) describe a large-scale restoration program being planned to reverse this wetland loss and the implications of both climate change and energy scarcity might have on that restoration. They argue that even more restoration might be necessary than formerly thought because of increased sea level rise coupled with changes in precipitation patterns

in the basin. They argue that less energy-intensive ecological engineering projects should be the approaches used to carry out the restoration in our energy-scarce future.

Mitsch et al. (2005b) continue to discussed wetland restoration in the Mississippi River Basin, but their restoration is proposed for improving water quality in the entire basin, particularly in the upper Midwest where excessive nutrients, particularly nitrates, are discharging from agricultural areas and causing a coastal eutrophication in the Gulf of Mexico thousands of kilometers away. This eutrophication, referred to as a hypoxia, can only be solved if nitrate-nitrogen fluxes are reduced substantially. They present an updated empirical model of nitrogen retention for wetlands receiving river or agricultural runoff calibrated from wetlands throughout the Mississippi River Basin. Some of the data that were used to construct the model are from the wetlands at the Olentangy River Wetland Research Park but additional data are from the Caernarvon diversion wetlands in Louisiana. In 10 years of operation, the experimental wetlands on the Olentangy River have removed 35% of the nitrate-nitrogen flowing through them, while the Caernarvon wetlands retained 55% nitrate-nitrogen for a similar loading rate. The nitrate retention model is in anticipation of the millions of hectares of wetlands and riparian ecosystems that need to be created and restored in the Basin to protect the Gulf of Mexico from hypoxia. They found that 22,000 km^2 of wetland creation and restoration in the Mississippi River Basin would be necessary to reduce the flux of nitrate-nitrogen to the Gulf by 40%.

Continuing on the topic of restoring the agricultural lands in the Mississippi River Basin, Hey et al. (2005) present an intriguing concept of "nitrogen farming" as a means to economically encourage agriculture in the Midwest to create wetlands and other methods to reduce nitrogen to the Gulf of Mexico. They argue that credit could be given to farmers for storing floodwaters as well as removing nitrogen, phosphorus, carbon, pesticides, sediments, and other constituents and that this trading could be done in the marketplace, lessening farmer's dependence on government subsidies while providing relieve for nonpoint source pollution and flooding. They suggest that removal of a metric ton of nitrogen in a wetland would be worth US$ 2500.

4.2. Long-term wetland restoration

We are now reaching the point where there are a few wetland restoration efforts have been followed for more than a few years and much can be learned from these long-term restorations about self-design and adaptive management. Two papers in this special issue discuss long-term wetland restoration projects of 30 years or more. Garbish (2005) presents a 30-year case history of a 0.8-ha salt marsh restoration in coastal Maryland, USA that he and his colleagues started in 1971. The initial restoration involved the movement of 3000 m^3 of sand, silt, and clay. Over 60,000 plants representing nine herbaceous species were introduced to the wetlands in 1972 with over 95% of the planted areas including *Spartina alterniflora*. A marsh "eatout" by Canada geese (*Branta canadensis*) in 1973 caused even more dominance by *S. alterniflora*. By 1992, 20 years after the wetland restoration began, a peat bank of 30 cm had developed throughout the marsh; this peat build-up had essentially kept up with sea level rise in the Chesapeake Bay. By 30 years after planting, the wetland area remained well established.

Brown (2005) describes large-scale landscape and wetland restoration in a 120,000 ha area of Florida dominated by phosphate mining that began in the state in the late 19th century. He describes the restoration as benefiting from adaptive management, ecological engineering, and recognition of the self-organizing capability of ecosystems and concludes that large-scale restoration needs to be done for ecosystems, not only certain vegetation strata, and that more recognition of the importance of early successional pioneer species is needed. He concludes that future research needs to address large-scale issues of how ecosystems need to be "hydrologically and ecologically organized" and how humans fit into this organization.

4.3. Creation of wetlands for mitigation of wetland loss

Three papers discuss wetland restoration and creation in light of the current "wetland trading" that is prevalent in the United States whereby wetland losses can be "mitigated" by wetland creation or restoration. The policy in the United States of "no net loss" means that when wetlands are destroyed for development, they are supposed to be replaced.

Brooks et al. (2005) argue that no net loss of wetland function has not occurred in the United States and present a model for comparing created and restored wetlands to reference wetlands. Their lab has identified 222 reference wetlands in five major ecoregions of Pennsylvania in the last decade. Data from these reference wetlands are used to both improve the design of new creation and restoration projects and to allow functional comparison with past mitigation projects. They found that created wetlands had lower organic matter and higher bulk density of soils and lower vegetation cover and plant richness than did reference wetlands.

A more optimistic picture is presented by Amon et al. (2005) who illustrate in a long-term experiment how it is possible to create fens in areas where hydric soils, which generally indicate the presence of previous wetlands, are not found. They conducted a multi-year experiment in glaciated Ohio on a $15\,m \times 32\,m$ site, where initially half of the site had hydric soil and the other half did not. Water was introduced to the plots and they were seeded with a mixture containing 33 species of fen plants. After 10 years of observation, it was concluded that fens can be constructed, they eventually develop characteristics similar to natural fens, new soils develop rapidly, and the initial presence of hydric soils is not necessary.

White and Fennessy (2005) illustrate a way in which a regional approach can be taken to find sites suitable for wetland creation and restoration when they are required by law for wetland loss mitigation. They illustrate a GIS model that predicts the suitability for wetland restoration in a $2100\,km^2$ watershed in northern Ohio, USA. The model uses criteria such as hydric soils, land use, topography, stream order, and saturation index. Using three versions of the GIS model, between 8 and 15% of the watershed was determined to have high potential for wetland restoration.

4.4. Conservation and restoration of the world's wetlands

Crisman et al. (2005) describe the restoration of two shallow lakes in northern Greece where fringe wetlands suggest a "horizontal" management scheme in addition to the normal "vertical" management used for deep lakes. They point out that shallow lakes are more controlled by "system memory" in the sediments than are deep lakes and that littoral vegetation zones and fringe wetlands strongly influence shallow lakes.

Junk (2005) describes the conservation, threats, and potential restoration of one of the great wetlands in the world, the $160,000\,km^2$ Pantanal in the center of South America. The Pantanal is in relatively good condition, supports a distinct biodiversity, and has a rich history of low-impact ranching and other light human changes. The Pantanal is relatively isolated from consumption centers in South America, still has relatively low population density and low-impact ranching, is still in relatively good health, and is managed by democratic governments. Recent interest in developing the area through a multi-nation *hidrovia* hydrologic modification project and the new world or intercontinental economics links have brought attention to how fragile the Pantanal is. Junk (2005) argues that there is an opportunity for careful development and protection of the wetlands and believes there is still time to protect the Pantanal by focusing the local economy on sustainable development, "green-label" products, and ecotourism.

Lewis (2005) provides a review of ecological engineering principals and practices for restoration of world's mangrove wetlands. Mangroves are found in $14,700\,km^2$ of the tropical shorelines of the world, down 26% from the estimated coverage of mangroves as recently as 1980. Largest losses are in the Philippines, Thailand, Vietnam, and Malaysia. Mangrove restoration projects often fail because "many mangrove restoration projects move immediately into planting of mangroves without determining why natural recovery has not occurred" or restoration is attempted where mangroves did not exist previously (Lewis, 2005). Millions of dollars have been wasted in failed mangrove restoration projects around the world. Lewis (2005) summarizes the most important principle in mangrove restoration and wetland restoration in general—determine the normal hydrology of natural mangroves in the region first. Getting the hydrology right is the first of seven mangrove restoration principles presented in his paper. Lewis (2005) concludes that "a common ecological engineering approach should be applied to habitat restoration projects" and that "an analytical thought process and less use of 'gardening' of mangroves" may be the solutions to mangrove restoration problems.

Acknowledgements

I appreciate the support given by the Ohio State University's Office of Research that made it possible for us to invite these distinguished wetland scientists to visit Columbus. I also wish to express publicly our appreciation to the many donors, students, and volunteers who collectively made the Olentangy River Wetland Research Park happen over the past 14 years. A special thanks to Dr. Jerry Pausch and Lenora Pausch for supporting the Moonlight on the Marsh lecture designations for all of these speakers. I also appreciate the great support that Dr. Li Zhang gave to the success of the Wetland Invitational and to OSU President Karen Holbrook and Vice-President Bob Moser for welcoming participants at this Wetland Invitational. Olentangy River Wetland Research Park Publication 05-002.

References

Amon, J.P., Jacobson, C.S., Shelley, M.L., 2005. Construction of fens with and without hydric soils. Ecol. Eng. 24, 341–357.

Brooks, R.P., Wardrop, D.H., Cole, C.A., Campbell, D.A., 2005. Are we purveyors of wetland homogeneity? A model of degradation and restoration to improve wetland mitigation performance. Ecol. Eng. 24, 331–340.

Brown, M.T., 2005. Landscape restoration following phosphate mining: thirty years of co-evolution of science, industry, and regulation. Ecol. Eng. 24, 309–329.

Crisman, T.L., Mitraki, C., Zalidis, G., 2005. Integrating vertical and horizontal approaches for management of shallow lakes and wetlands. Ecol. Eng. 24, 379–389.

Day Jr., J.W., Barras, J., Clairain, E., Johnston, J., Justic, D., Kemp, G.P., Ko, J.-Y., Lane, R., Mitsch, W.J., Steyer, G., Templet, P., Yañez-Arancibia, A., 2005. Implications of global climatic change and energy cost and availability for the restoration of the Mississippi Delta. Ecol. Eng. 24, 253–265.

Garbish, E.W., 2005. Hambleton Island restoration: environmental concern's first wetland creation project. Ecol. Eng. 24, 289–307.

Hey, D.L., Urban, L.S., Kostel, J., 2005. Nutrient farming: the business of environmental management. Ecol. Eng. 24, 279–287.

Junk, W.J., 2005. Pantanal: a large South American wetland at a crossroads. Ecol. Eng. 24, 391–401.

Lewis, R.R., 2005. Ecological engineering for successful management and restoration of mangrove forests. Ecol. Eng. 24, 403–418.

Mississippi River/Gulf of Mexico Watershed Nutrient Task Force, 2001. Action plan for reducing, mitigating, and controlling hypoxia in the northern Gulf of Mexico. Report submitted to U.S. Congress, U.S. Environmental Protection Agency, Washington, DC.

Mitsch, W.J., Wu, X., Nairn, R.W., Weihe, P.E., Wang, N., Deal, R., Boucher, C.E., 1998. Creating and restoring wetlands: A whole-ecosystem experiment in self-design. BioScience 48, 1019–1030.

Mitsch, W.J., Day Jr., J.W., Gilliam, J.W., Groffman, P.M., Hey, D.L., Randall, G.W., Wang, N., 2001. Reducing nitrogen loading to the Gulf of Mexico from the Mississippi River Basin: strategies to counter a persistent large-scale ecological problem. BioScience 51, 373–388.

Mitsch, W.J., Wang, N., Zhang, L., Deal, R., Wu, X., Zuwerink, A., 2005a. Using ecological indicators in a whole-ecosystem wetland experiment. In: Jørgensen, S.E., Xu, F.-L., Costanza, R. (Eds.), Handbook of Ecological Indicators for Assessment of Ecosystem Health. CRC Press, Boca Raton, FL, pp. 211–235.

Mitsch, W.J., Day Jr., J.W., Zhang, L., Lane, R., 2005b. Nitrate-nitrogen retention by wetlands in the Mississippi River Basin. Ecol. Eng. 24, 267–278.

National Research Council, 2001. Compensating for Wetland Losses under the Clean Water Act. National Academy Press, Washington, DC, 322 pp.

White, D., Fennessy, S., 2005. Modeling the suitability of wetland restoration at the watershed scale. Ecol. Eng. 24, 359–377.

Available online at www.sciencedirect.com

SCIENCE DIRECT°

Ecological Engineering 24 (2005) 253–265

ELSEVIER

ECOLOGICAL ENGINEERING

www.elsevier.com/locate/ecoleng

Implications of global climatic change and energy cost and availability for the restoration of the Mississippi delta

John W. Day Jr. [a,b,*], John Barras [d], Ellis Clairain [e], James Johnston [d],
Dubravko Justic [a,b], G. Paul Kemp [b], Jae-Young Ko [b], Robert Lane [b],
William J. Mitsch [f], Gregory Steyer [d], Paul Templet [c], Alejandro Yañez-Arancibia [g]

[a] *Department of Oceanography and Coastal Sciences, School of the Coast and Environment,
Louisiana State University, Baton Rouge, LA 70803, USA*
[b] *Coastal Ecology Institute, School of the Coast and Environment, Louisiana State University, Baton Rouge, LA 70803, USA*
[c] *Department of Environmental Studies, School of the Coast and Environment, Louisiana State University, Baton Rouge, LA 70803, USA*
[d] *U.S. Geological Survey, National Wetlands Research Center, Baton Rouge, LA 70894, USA*
[e] *Engineer Research and Development Center, U.S. Army Corps of Engineers, Vicksburg, MS 39180, USA*
[f] *Olentangy River Wetland Research Park, School of Natural Resources, Ohio State University,
352 W. Dodridge Street, Columbus, OH 43202, USA*
[g] *Coastal Ecosystems Unit, Institute of Ecology A.C., Km 2.5 Carretera Antigua Xalapa-Coatepec No. 351,
El Haya 91070, Xalapa Ver., México*

Received 9 July 2004; received in revised form 28 October 2004; accepted 1 November 2004

Abstract

Over the past several thousand years, inputs from the Mississippi River formed the Mississippi delta, an area of about 25,000 km². Over the past century, however, there has been a high loss of coastal wetlands of about 4800 km². The main causes of this loss are the near complete isolation of the river from the delta, mostly due to the construction of flood control levees, and pervasive hydrological disruption of the deltaic plain. There is presently a large-scale State-Federal program to restore the delta that includes construction of water control structures in the flood control levees to divert river water into deteriorating wetlands and pumping of dredged sediment, often for long distances, for marsh creation. Global climate change and decreasing availability and increasing cost of energy are likely to have important implications for delta restoration. Coastal restoration efforts will have to be more intensive to offset the impacts of climate change including accelerated sea level rise and changes in precipitation patterns. Future coastal restoration efforts should also focus on less energy-intensive, ecologically engineered management techniques that use the energies of nature as much as possible. Diversions may be as important for controlling salinity as for providing sediments and nutrients for restoring coastal wetlands. Energy-intensive pumping-dredged sediments for coastal restoration will likely become much more expensive in the future.
© 2005 Elsevier B.V. All rights reserved.

Keywords: Mississippi delta; Salinity intrusion; Climate change; Energy

---

* Corresponding author.
E-mail address: johnday@lsu.edu (J.W. Day Jr.).

0925-8574/$ – see front matter © 2005 Elsevier B.V. All rights reserved.
doi:10.1016/j.ecoleng.2004.11.015

1. Introduction

The Mississippi River has the largest discharge and drainage basin in North America and is one of the largest rivers in the world. The watershed encompasses about 3 million km², about 40% of the area of the lower 48 United States, and accounts for about 90% of the freshwater inflow to the Gulf of Mexico. The Mississippi delta is the largest contiguous coastal ecosystem in the U.S.; approximately 60% of the estuaries and marshes in the Gulf of Mexico are located in coastal Louisiana. The delta is ecologically and economically important. Ecologically, the coastal wetlands and shallow waters of the delta provide habitat for fish and wildlife, produce food, regulate chemical transformations, maintain water quality, store and release water, and buffer storm energy (Day et al., 1997, 2000). These processes support a variety of economically important, natural resource-based activities valued at several billion dollars annually including recreational and commercial fisheries, fur mammals, and alligators, ecotourism, and hunting (LCWCRTF, 1993; Day et al., 1997). The lower Mississippi River in Louisiana is home to the largest port activity by tonnage in the world

(LCWCRTF, 1993). Petroleum products produced by refineries located in the Louisiana coastal zone are valued at US$ 30 billion annually and approximately 20% of crude oil and 33% of natural gas of the United States flow through the Louisiana coastal zone (Davis and Guidry, 1996).

The Mississippi delta (Fig. 1) was formed over the past 6000–7000 years as a series of overlapping delta lobes (Roberts, 1997). There was an increase in wetland area in active deltaic lobes and wetland loss in abandoned lobes, but there was an overall net increase in the area of wetlands over the past several thousand years. Currently, only two of the distributaries of the river are functioning, the lower river and the Atchafalaya River which carries about one-third of the total flow of the Mississippi River. At the time European occupation began, however, numerous distributaries were functioning, either year round or during the seasonal spring flood. The delta was sustained by a series of energetic forcings or pulsing events that occurred over different spatial and temporal scales. These pulses include shifting deltaic lobes, crevasses, great river floods, hurricanes, annual river floods, frontal passages, and tides (Day et al., 1997, 2000). The area of the delta is about

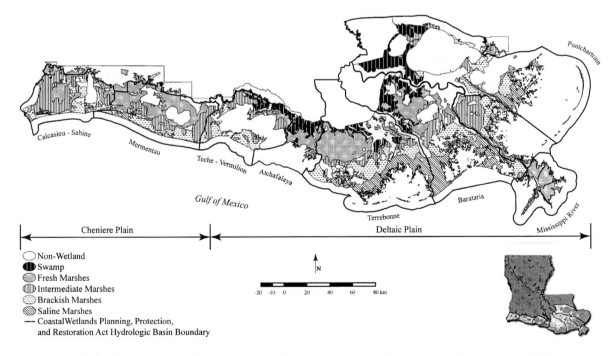

Fig. 1. Louisiana coastal zone showing estuarine basins and vegetation zones (modified from Lindscombe et al., 2001).

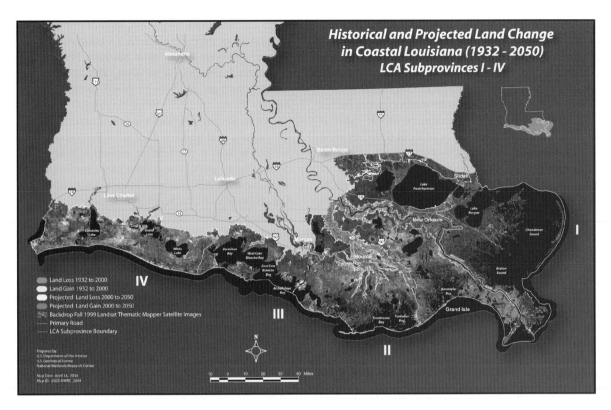

Fig. 2. Historical and projected land change trend in coastal Louisiana, 1932–2050 (from U.S. Geological Survey, National Wetlands Research Center).

25,000 km² including wetlands, shallow inshore water bodies, and low elevation uplands (mostly associated with river distributary ridges and beach ridges).

There was, however, a dramatic reversal of this condition of net growth in the 20th century. There has been an enormous loss of coastal lands in the delta with a total loss of about 4800 km² since the 1930s (Fig. 2; Boesch et al., 1994; Britsch and Dunbar, 1993; Barras et al., 1994, 2003). Over 95% of this loss was wetland, primarily as marshes converted to open water. In the 1970s, the loss rate was as high as 100 km²/year and the loss rate from 1990 to 2000 was about 60 km²/year (Barras et al., 2003). Between 1956 and 2000, the average loss rate was 88 km²/year (Barras et al., 1994, 2003). These high rates of wetland loss are projected to continue for the next half century; and by 2050, it is estimated there will be an additional net wetland loss of 1329 km² (Barras et al., 2003).

An understanding of the causes of this land loss is important not only for a scientific comprehension of the

mechanisms involved, but also so that effective management plans can be developed to restore the delta (see Boesch et al., 1994; Day et al., 2000 for a review of these issues). A number of factors led to the massive loss of wetlands. Foremost among these are flood-control levees along the Mississippi River that resulted in the elimination of riverine input to most of the delta (Boesch et al., 1994; Day et al., 2000). In addition to the flood-control levees, most active distributaries were closed, and the river mouth was made more efficient for navigation by dredging. This resulted in the loss of most river sediments, which once sustained the wetlands, directly to deep waters of the Gulf of Mexico. There has also been a reduction of the suspended sediment load in the Mississippi River caused by dam construction in the Upper Mississippi River (Kesel, 1988, 1989).

Pervasive altered wetland hydrology, mostly caused by canals, is another important factor contributing to wetland loss. Canals, originally dredged for drainage and navigation, are now overwhelmingly linked to the

petroleum industry. Drilling access canals, pipeline canals, and deep-draft navigation channels have left a dense network of about 15,000 km of canals in the coastal wetlands. Although canals are estimated to comprise about 2.5% of the total coastal surface area, their destructive impact has been much greater (Turner et al., 1982). Spoil banks, composed of the material dredged from the canals, interrupt sheet flow, impound water, and cause deterioration of marshes. Long, deep navigation canals that connect saline and freshwater areas tend to lessen freshwater retention time, and allow greater inland penetration of saltwater.

In sum, there is a broad consensus that wetland loss is a complex interaction of a number of factors acting at different spatial and temporal scales (e.g., Turner and Cahoon, 1987; Day and Templet, 1989; Boesch et al., 1994; Day et al., 1995, 1997). Day et al. (2000) concluded that isolation of the delta from the river by levees was perhaps the most important factor.

2. Approaches to restoration of the Mississippi delta

The State of Louisiana and the Federal government have embarked on an ambitious program, called the Louisiana Coastal Area (LCA) Ecosystem Restoration Plan, to restore the Mississippi delta (http://www.lca.gov). In the remainder of this paper, we describe elements of the restoration program and discuss the implications for restoration of two global trends, climate change and the cost and availability of energy.

The primary approaches to restoration of the delta involve reversal of the impacts that have occurred due to human activity. These include the near complete separation of the river from the delta and the pervasive hydrological alteration of the delta plain. There are several primary approaches to addressing these impact including river diversions, hydrological restoration, marsh creation and restoration using dredged sediments, and barrier island restoration.

2.1. Diversions of river water into deteriorating wetlands

Freshwater diversions are the major management approach for restoration of the Mississippi delta

(Fig. 3a). There are at present, two diversions operating (with a maximum discharge between 200 and $250 \, m^3 \, s^{-1}$) with several more planned. The aim is to reconnect the river with its delta. The idea of diversions is not new and was proposed nearly a century ago by Viosca in 1927. When freshwater diversions were first planned in Louisiana over three decades ago, the primary goal was to reduce salinity to enhance oyster production in surrounding regions (Chatry et al., 1983; Chatry and Chew, 1985). More recently, diversions have increasingly been used as a way of delivering sediments and nutrients to wetlands in an attempt to counter relative-sea-level-rise (RSLR) and restore deteriorating wetlands (Day and Templet, 1989; Day et al., 1997, 2000; LCA Plan, http://www.lca.gov). Recently completed and ongoing research indicates that diversions lead to enhanced accretion, substantial reductions in nutrients, higher marsh productivity, and higher fishery yield (Day et al., 1997; Lane et al., 1999, 2001, 2004; Perez et al., 2003; DeLaune and Pezeshki, 2003; DeLaune et al., 2003). There is concern that nutrients in diverted water will lead to eutrophication and there is continuing research on the issue and ways to manage diversions to minimize the potential for water quality problems.

There is concern that introductions of river water will lead to water quality problems in estuaries and coastal bays similar to the hypoxia zone in the Gulf of Mexico. This zone forms when excess nitrogen in the river, mainly from agricultural runoff, lead to algal blooms that sink and consume oxygen (Rabalais et al., 1996). It is thus imperative that restoration of the Mississippi basin be carried out at the same time as delta restoration. Mitsch et al. (2001) suggested an ecological engineering approach for the basin where constructed and restored wetlands are used to reduce nutrients in agricultural runoff.

2.2. Reopening distributaries and hydrologic restoration

In addition to diversions, the LCA planning process is studying the potential for reopening one or more of the Mississippi River distributaries that have been closed (i.e., Bayou Lafourche) and putting more water down the Atchafalaya River. All of these efforts are designed to introduce river water into shallow inshore areas to restore coastal wetlands.

Fig. 3. (a) A river diversion at Caernarvon, LA. The gated structure allows Mississippi River water to flow into the adjacent coastal system (Courtesy of U.S. Army Corps of Engineers, New Orleans District). (b) A typical sediment dredging operation in a coastal bay in Louisiana (Courtesy of U.S. Army Corps of Engineers, New Orleans District).

Another important aspect of coastal restoration is hydrologic restoration. As stated earlier, the pervasive hydrologic disruption of the delta has led to reductions in overland flow, decreased sedimentation, changes in wetland productivity, increased flooding, higher subsidence, and salt water intrusion (Swenson and Turner, 1987; Reed, 1992; Cahoon, 1994; Boumans and Day, 1994; Conner and Day, 1988, 1991, 1992; Morton et al., 2002). Hydrological restoration seeks

to reduce these impacts by such management activities as spoil bank removal, closure of some deep navigation channels (such as the Mississippi River Gulf Outlet, southeast of New Orleans) and putting locks in others (Day and Templet, 1989; Turner and Streever, 2002; Day et al., 2004). Such restoration can be particularly effective if done in conjunction with diversions so that river water is used most effectively.

2.3. Use of dredged sediments for wetlands creation and restoration

Dredged sediments have long been used for wetland creation (Fig. 3b). This has been done rather extensively in the "bird's foot" delta at the mouth of the river and in the Atchafalaya delta. Dredged sediments have also been used in many parts of the coast where dredging projects have been carried out. However such use of dredged sediments as part of navigation projects has been done opportunistically and can address only a small part of coastal land loss. Recent work has shown that nourishment of low elevation, unhealthy marshes with dredged sediments results in an increase in marsh elevation and thus productivity (Mendelssohn and Kuhn, 2003). There are now plans to dredge sediments specifically for marsh creation and nourishment and to pump the sediments over long distances (10s of km). Such pumping is expensive, economically and energetically, and costs increase as boosters and pipe are added to pump longer distances. Currently the average cost of sediment dredging for wetlands restoration in the northern Gulf of Mexico is about US$ 40,000/ha, excluding additional activities such as construction of protective structures, planting, re-contouring, and monitoring (Turner and Streever, 2002, p. 95).

2.4. Barrier island restoration

Barrier islands in coastal Louisiana are important for protecting coastal wetlands by mitigating wave energy, to provide wildlife habitat for migrant birds and to maintain estuarine conditions. However, the islands have eroded significantly, due to RSLR, reduced sediment availability, and erosion during storms (Dingler et al., 1992). Hurricanes are particularly damaging to barrier islands (Stone et al., 1997). Barrier islands are restored by pumping sands from offshore to rebuild the islands. At times, engineered structures are also used. Sand-trapping fences and vegetative plantings are used to stabilize sand dunes on barrier islands along with beach nourishment (LA DNR, 1997, p. 15). After vegetative planting, re-vegetation takes place and is affected by both biotic and abiotic factors (e.g., soil salinity, Courtemanche et al., 1999). Control of grazers, such as nutria, is also important to maintaining vegetative cover (Hester et al., 1994).

Barrier island restoration is expensive. For example, the cost for 1700 liner feet of beach front restoration at East Island, Louisiana, was US$ 1.3 million (LA DNR, news release, April 18, 1996), and around US$ 500 million are needed to restore about 85 miles of barrier islands in coastal Louisiana (St. Pé, 1999). Barrier islands must be renourished periodically to ensure sustainability.

3. Global trends and their implications for delta restoration

3.1. Global climate change

There is a broad consensus in the scientific community that human activity is affecting global climate (IPCC, 2001). Climate change will significantly alter many of the world's coastal and wetland ecosystems (Poff et al., 2002). Global climate change will affect temperature, the amount and seasonality of rainfall, and the rate of sea level rise. The Intergovernmental Panel on Climate Change (IPCC) predicts that global temperatures will rise from 1 to 5 °C during the 21st century. This increase in temperature will affect coastal biota directly and lead to changes in precipitation and an acceleration of sea level rise. It is predicted that as the tropics gain more heat, there will be a greater transport of water vapor toward higher latitudes. Thus, it is likely that, in general, lower latitudes will experience a decrease in precipitation and higher latitudes will experience an increase in rainfall.

General circulation models (GCMs) are not consistent in their predictions of the effects of climate change on precipitation and temperature, which are two important drivers of freshwater inflow to estuaries. For example, runoff estimates for the Mississippi River basin differ greatly between the Canadian CGCM1 model and the Hadley HADCM2 model (Wolock and McCabe, 1999). Both models predict an increase in future extreme rainfall and runoff events, but they disagree in terms of both the magnitude and direction of changes in average annual runoff. The average annual runoff of the Mississippi River basin, for example, was projected to decrease by 30% for the Canadian model, but increase by 40% for the Hadley model by the year 2099. Estimated changes of freshwater inflow into major U.S. estuaries projected by the Hadley model by

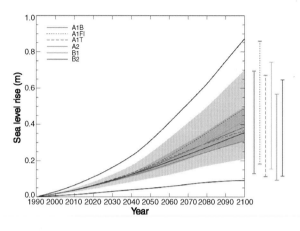

Fig. 4. Projected global average sea level rise, 1990–2100 (*Source:* IPCC, 2001).

the year 2099 range from −40% to +100%. Similar calculations based on the Canadian model indicate significantly reduced inflows for all coastal regions except the U.S. Pacific coast (Wolock and McCabe, 1999). It is likely many coastal and estuarine ecosystems will experience changes in freshwater inflow. However, at present it is unclear in what manner these changes will occur. There has already been a small, but significant, increase in the flow of the river during the past half-century (Justic et al., 2003). While there is much uncertainty in predictions of local changes in precipitation and runoff, the precautionary principle suggests that management plans for delta should take such changes into consideration.

There is strong consensus that global warming will lead to accelerated eustatic sea level rise in the 21st century. IPCC (2001) predicted that sea level will rise by 20–65 cm in the 21st century, with a best estimate of 30–50 cm (Fig. 4). This is much higher than measured eustatic sea level rise for the 20th century of 1–2 mm/year (Gornitz et al., 1982). This increase in sea level must be added to subsidence to obtain the RSLR that coastal wetlands in the Mississippi delta will be subject to over the 21st century. Thus, RSLR in the delta will increase from about 1 cm/year to 1.3–1.7 cm/year within this century (a 30–70% increase).

Two important physiological reasons that lead to the loss of wetlands are flooding stress due to increased flooding duration and salinity stress (Mendelssohn and Morris, 2000). Global climate change will likely exacerbate both of these stresses. Accelerated sea level rise will lead to significant increase in flooding duration. Unless wetlands can accrete vertically at the same rate as water level rise, coastal vegetation will become progressively more stressed and ultimately die. Even at current rates of RSLR of about 1 cm/year, most wetlands of the Mississippi delta do not have sufficient rates of vertical accretion to survive (Delaune et al., 1983; Hatton et al., 1983; Conner and Day, 1991). Rising sea level combined with lower freshwater input will lead to increased saltwater intrusion and salinity stress. This especially threatens the extensive tidal freshwater wetlands of the delta. This combination of high RSLR, increased temperature, and lower freshwater input results in the north central Gulf having the highest vulnerability to climate change in the United States (Thieler and Hammar-Klose, 2001).

Increased temperature may interact with other stressors to damage coastal marshes. For example, during the spring to fall period of 2000 in the Mississippi delta, there were large areas of salt marsh that were stressed and dying. This appears to be the result of a combination of effects related to a strong la Niña event which resulted in sustained low water levels caused by a global circulation pattern, prolonged and extreme drought and high air temperatures. This combination of factors apparently raised soil salinities to stressful and even toxic levels. McKee et al. (2004) suggested that increases in temperature and decreases in rainfall associated with climate change may dramatically affect tidal marshes.

Another major climate driver in warm temperate zones is reduction in the frequency of extreme freezes, which is presently occurring along the coastal fringes of the Mississippi River Deltaic Plain. An important result of increasing temperature along the northern Gulf of Mexico will likely be a northward migration of mangroves replacing salt marshes. Mangroves are tropical coastal forests that are freeze-intolerant. Chen and Twilley (1998) developed a model of mangrove response to freeze frequency. They found that when freezes occurred more often than once every 8 years, mangrove forests could not survive. At a freeze frequency of 12 years, mangroves replaced salt marsh. Along the Louisiana coast, freezes historically occurred about every four years. By the spring of 2004, however, a killing freeze had not occurred for 15 years and small mangroves occur over a large area near the coast. If this trend continues, mangroves will probably spread over much of the northern Gulf and part

of the south Atlantic coast. In fact, mangroves are already becoming established and more widespread due to warming. According to Dr. J. Visser (personal communication, Coastal Ecology Institute, LSU, June 25, 2004), there are approximately 150 km^2 of mangrove habitat in the Barataria and Terrebonne basins. It was observed during the drought in 2000, that mangroves were able to withstand higher temperatures and salinity and water stress (McKee et al., 2004). Thus, succession from tidal marshes to mangroves may lessen the impact of drier, higher salinity conditions.

Because mangroves have many of the same ecological functions as salt marshes (high productivity, habitat for wildlife and fishes, sites of nutrient uptake, etc.), a switch in the U.S. coastal wetlands from salt marshes to mangroves might not change ecosystem function much. However, if the climate becomes more variable with freeze-free periods interspersed with occasional hard freezes, it could be more difficult for either marshes or mangroves to survive, resulting in a loss of wetland habitat.

3.2. The availability and cost of energy

The availability and cost of energy will likely become an important factor affecting the way that natural resource management is carried out in the future. Over the past decade, increasing information has appeared in the scientific literature suggesting that world oil production will peak within a decade or two (Fig. 5a) implying that demand will consistently be greater than supply and that the cost of energy will increase significantly in the coming decades. This information has come primarily from petroleum geologists with long experience in petroleum production (Masters et al., 1991; Campbell and Laherrère, 1998; Kerr, 1998; Bentley, 2002; Deffeyes, 2001, 2002; Hall et al., 2003; Heinberg, 2003).

The various projections of when world oil production will peak are based on the approach developed by M. King Hubbert who became well known because of his famous prediction made in 1956 that U.S. oil production would peak in the early 1970s (it peaked in 1971). Hubbert also predicted that world oil production would peak around the year 2000 (see Deffeyes, 2001 and Heinberg, 2003 for a discussion of Hubbert's work). In essence, projections of future oil production and peak oil production use statistical and physical

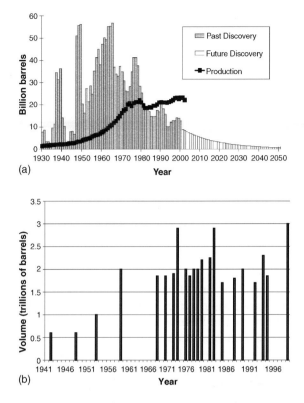

Fig. 5. (a) Historical global oil production and past and projected oil discovery, 1930–2050 (*Source:* Campbell, 2004; http://www.oilcrisis.com/campbell/). (b) Published estimates of total ultimate oil recovery (modified from Hall et al., 2003, p. 321).

methods based on reserve estimates and the lifetime production profile of typical oil reservoirs. Oil production from reservoirs tends to follow a bell-shaped curve with a rapid increase in production followed by a relatively rapid decrease in production. Thus, by knowing the early production history of a reservoir (or many reservoirs together) and an estimate of reserves, the time of peak production and total oil production can be estimated. Based on this information, various authors have predicted that world oil production will peak sometime during the first two decades of the 21st century (see Masters et al., 1991; Campbell and Laherrère, 1998; Deffeyes, 2001; Bentley, 2002; Heinberg, 2003). Some have augured that additional discoveries will provide abundant oil well into the future (see Deffeyes, 2001; Hall et al., 2003; Heinberg, 2003 for a review). But most estimates of ultimately recoverable oil (URO) have remained relatively constant since about 1965 at about 2 trillion barrels (Fig. 5b). This is the value of

URO that most authors have used to predict the timing of peak oil production. Oil discoveries peaked in the 1950s and 1960s and have declined substantially since (Fig. 5a). We now consume about two barrels of oil for each one discovered.

An important factor that affects consideration of energy use is energy return on investment (EROI). EROI is the ratio of the energy in oil that is produced to all the energy used to discover and produce oil. During the period of exponential increases in conventional oil production, the EROI was between 100:1 and 50:1. Over the last two decades, EROI for world conventional oil production fell to between 20:1 and 10:1 (Cleveland et al., 1984; Cleveland, 2005). Thus, it is costing more and more to find and produce oil. The EROI for nonconventional sources of oil (oil shale and oil sands) and most renewables are all less than 15:1 and most are substantially less than 10:1 (Heinberg, 2003). We believe that the implication of all of this is that oil production will not be able to meet demand and that the cost of energy will increase substantially.

4. Implications for delta restoration and management

Global climate change and energy availability have important implications for management of the Mississippi delta. Global climate change is predicted to lead to accelerated sea level rise, increased temperatures, perhaps higher Mississippi River discharge, but lower precipitation and local freshwater discharge to Louisiana estuaries. Droughts, such as occurred in 2000, may become more frequent and salt-water intrusion will likely be more severe. Thus, two of the most detrimental stressors leading to vegetation mortality, prolonged flooding and salinity, probably will be exacerbated with climate change. Vertical accretion rates will have to be 30–70% higher in a century than at present if coastal marshes are to survive. Both river diversions and utilization of dredged sediment will become even more important in the future in enabling coastal marshes to survive accelerated sea level rise. River water introduces sediments that contribute directly to vertical accretion, nutrients that stimulate vegetation productivity and organic soil formation, and iron that detoxifies sulfide by precipitation (DeLaune and Pezeshki, 2003; DeLaune et al., 2003). An important consideration is that marshes cre-

ated with dredged sediments will have a shorter lifetime in the future than at present because of the acceleration of the rate of sea level rise. It is a paradox that river discharge is predicted to increase significantly by one GCM while local freshwater discharge is predicted to decrease. Therefore, one of the most important benefits of river diversions will probably be the use of the excess water in the Mississippi River to counter higher salinities due to sea level rise and lower local fresh water surpluses. Thus, river diversions will probably become more critical in delta restoration to offset the impacts of climate change (Table 1).

Energy price and availability will also likely have important implications for delta restoration. The remarkable economic growth of the last century has been fuelled by cheap energy, most importantly oil (Hall et al., 2003). Much of the infrastructure of our industrial society was put in place using 50:1 to 100:1 oil (and equivalently cheap coal and natural gas). The massive flood control and navigation works on the Mississippi River were part of this infrastructure. This includes the giant Bonnet Carré Spillway built on the Mississippi River in the early 1930s to protect New Orleans. The spillway is still functioning today.

If the end of era of cheap energy comes to pass, energy intensive management methods will become increasingly expensive and untenable. Louisiana has relatively little political power at the national level, so it is possible that we will not be able to obtain large amounts of resources for delta restoration. We will be forced to consider less energy-intensive, less expensive options for restoring deteriorating coastal marshes in the post-oil peak era.

Increased cost and reduced availability of energy suggests that those methods of restoration that use relatively low amounts of energy are the ones most sustainable in the long term. The combination of diversions with use of dredged sediments is very important because marshes created with dredged sediments will have a shorter lifetime in the future because of increased sea level rise. Restoration of the Mississippi basin should also be carried out in conjunction with delta restoration in order to minimize water quality problems as well as solve environmental problems throughout the basin (Mitsch et al., 2001; Day et al., 2003). Pumping sediments, which is very expensive energetically, has a role early in the restoration program in conjunction with construction of diversions.

Table 1
Approaches to restoration of the Mississippi delta

Approach	Description	Effects	Cost
Freshwater diversion	Diverting Mississippi River water into coastal wetlands for the purpose of restoration using siphons or gates. This is ecological engineering. A major restoration tool	Dissolved nutrients, freshwater, and suspended sediments contribute the nourishment of degraded marshes, along with enhanced fishery habitat. Impacts are ecosystem wide, rather than specific or individual. Land building is a long-term goal. Large scale	Relatively high capital cost, and low annual operation and maintenance cost
Hydrologic restoration	Restoring natural drainage patterns to remedy altered hydrology. Includes backfilling of canals, closure of canals, restoration of natural drainage features, reducing salt water intrusion. A secondary restoration tool	The integrity of altered wetlands is restored by a return towards natural hydrology. Small scale	Moderate to high financial and energy costs per unit area
Use of dredged sediments for wetlands creation/restoration	Beneficial use of sediment materials that can come from dredging projects or that are dredged specifically for the purpose of restoration. A major restoration tool	Newly created intertidal flats can be rapidly colonized by vegetation or planted. Most effective when used in conjunction with diversions	High financial and energy costs. Periodic additional sediment may be required
Barrier island restoration	Deposition of coarse dredged materials to increase barrier island height and width. Engineering structures, sand trapping, and vegetative plantings are used to stabilize sand dunes on barrier islands. A major restoration tool	Rapid restoration. Requires perpetual nourishment to maintain barrier islands. Barrier island restoration restores island habitat, protects wetlands, and maintains estuarine conditions	Labor intensive, capital intensive, very costly (e.g., US$ 1.3 million for 500 m of beach front restoration at East Island, Louisiana)

Restoration project type descriptions for CWPPRA-funded restoration projects in coastal Louisiana (LA DNR, 1997).

In this way, wetlands created can be nourished and sustained with diversions that are likely to have a life of a century or more.

Wherever possible, gravity, winds, and tides should be used to move water and sediments. Diversions, such as the Bonnet Carré Spillway and the diversion at Caernarvon, Louisiana, are examples of this. The Bonnet Carré was constructed over 70 years ago; it is still functioning and is likely to do so for decades. This argues for the construction of needed diversion structures in the near future when energy is still relatively cheap. After construction, they can be operated using relatively little energy, probably for the next century and longer. Energy intensive approaches, such as pumping sediments over long distances to build wetlands, should be used now but they may not be affordable 40–50 years from now. New land created by pumping sediments will last for a shorter period than would have been the case in the past because of the acceleration of sea level rise and higher salinities. Thus, diversions of river wa-

ter should be planned in conjunction with pumping to maintain the marshes. Modelling has shown that that riverine input can create and maintain wetlands (Martin et al., 2002).

This approach of using the energies of nature to the greatest extent possible is called ecological engineering. This is the ecological principle where small amounts of fossil fuel energies are used to channel much larger flows of natural energies (Odum, 1971; Mitsch and Jørgensen, 2003). Ecological engineering offers both a conceptual and practicable approach for long-term management of the delta in an era when the cost of fossil energies will become much more expensive.

In conclusion, we have presented evidence that climate and the availability and cost of energy will change significantly in the 21st century. Such changes have major implications for the restoration of the Mississippi delta. However, since the future is uncertain, this is a cautionary tale. But we believe that it is unwise to not

take these factors into consideration because choices made will strongly affect future restoration activities in the delta.

Acknowledgements

Support for this effort was partially provided by National Oceanic and Atmospheric Administration through the Louisiana State Grant College Program (NOAA Grant No. NA16G2249). Additional support was provided by the Corps of Engineers and the U.S. Geological Survey.

References

Barras, J.A., Bourgeois, P.E., Handley, L.R., 1994. Land loss in coastal Louisiana, 1956–1990. National Biological Survey, 10 color prints. National Wetlands Research Center Open File Report 94-01, 4 pp.

Barras, J.A., Beville, S., Britsch, D., Hartley, S., Hawes, S., Johnston, J., Kemp, P., Kinler, Q., Martucci, A., Porthouse, J., Reed, D., Roy, K., Sapkota, S., Suhayda, J., 2003. Historical and projected coastal Louisiana land changes: 1978–2050. USGS Open File Report 03-334, 39 pp.

Bentley, R.W., 2002. Global oil and gas depletion: an overview. Energy Policy 30, 189–205.

Boesch, D.F., Josselyn, M.N., Mehta, A.J., Morris, J.T., Nuttle, W.K., Simenstad, C.A., Swift, D., 1994. Scientific assessment of coastal wetland loss, restoration and management in Louisiana. J. Coastal Res. 20, 1–103 (Special issue).

Boumans, R.M., Day Jr., J.W., 1994. Effects of two Louisiana's marsh management plans on water and materials flux and short-term sedimentation. Wetlands 14 (4), 247–261.

Britsch, L., Dunbar, J., 1993. Land loss rates: Louisiana coastal plain. J. Coastal Res. 9, 324–338.

Cahoon, D., 1994. Recent accretion in two managed marsh impoundments in coastal Louisiana. Ecol. Appl. 4, 166–176.

Campbell, C.J., 2004. The growing gap (http://www.oilcrisis.com/campbell/).

Campbell, C.J., Laherrère, J.H., 1998. The end of cheap oil. Sci. Am., 60–65.

Chatry, M., Dugas, R.J., Easley, K.A., 1983. Optimum salinity regime for oyster production on Louisiana's state seed grounds. Contrib. Mar. Sci. 26, 81–94.

Chatry, M., Chew, D., 1985. Freshwater diversion in coastal Louisiana: recommendations for development of management criteria. In: Proceedings of the Fourth Coastal Marsh and Estuary Management Symposium, pp. 71–84.

Chen, R.H., Twilley, R.R., 1998. A gap dynamic model of mangrove forest development along gradients of soil salinity and nutrient resources. J. Ecol. 86, 37–51.

Cleveland, C.J., Costanza, R., Hall, C.A.S., Kaufmann, R., 1984. Energy and the U.S. economy: a biophysical perspective. Science 225, 890–897.

Cleveland, C.J., 2005. Net energy from the extraction of oil and gas in the United States Energy 30, 769–782.

Conner, W.H., Day Jr., J.W., 1988. Rising water levels in coastal Louisiana: importance to forested wetlands. J. Coastal Res. 4, 589–596.

Conner, W.H., Day Jr., J.W., 1991. Variations in vertical accretion in a Louisiana swamp. J. Coastal Res. 7, 617–622.

Conner, W.H., Day Jr., J.W., 1992. Water level variability and litterfall productivity of forested freshwater wetlands in Louisiana. Am. Midl. Naturalist 128, 237–245.

Courtemanche, R.P., Hester, M.W., Mendelssohn, I.A., 1999. Recovery of a Louisiana barrier island marsh plant community following extensive hurricane-induced overwash. J. Coastal Res. 15, 872–883.

Davis, D., Guidry, R.J., 1996. Oil, oil spills and the state's responsibilities. Basin Res. Inst. Bull. 6, 60–68.

Day, J.W., Pont, D., Hensel, P.F., Ibanez, C., 1995. Impacts of sea-level rise on deltas in the Gulf of Mexico and the Mediterranean: the importance of pulsing events to sustainability. Estuaries 18, 636–647.

Day Jr., J.W., Martin, J.F., Cardoch, L., Templet, P.H., 1997. System functioning as a basis for sustainable management of deltaic ecosystems. Coastal Manage. 25, 115–153.

Day Jr., J.W., Templet, P.H., 1989. Consequences of sea level rise: implications from the Mississippi delta. Coastal Manage. 17, 241–257.

Day Jr., J.W., Britsch, L.D., Hawes, S.R., Shaffer, G.P., Reed, D.J., Cahoon, D., 2000. Pattern and process of land loss in the Mississippi delta: a spatial and temporal analysis of wetland habitat change. Estuaries 23, 425–438.

Day Jr., J.W., Yañéz Arancibia, A., Mitsch, W.J., Lara-Dominguez, A.L., Day, J.N., Ko, J.-Y., Lane, R., Lindsey, J., 2003. Using ecotechnology to address water quality and wetland habitat loss problems in the Mississippi basin: a hierarchical approach. Biotechnol. Adv. 22, 135–159.

Day, J., Templet, P.H., Ko, J.-Y., et al., 2004. The Mississippi delta: system functioning, environmental impacts, and sustainable management. In: Caso, M. (Ed.), Environmental Diagnosis of the Gulf of Mexico, vol. 2. Mexican National Institute of Ecology, Mexico City, Mexico, pp. 851–880.

Deffeyes, K.S., 2002. World's oil production peak reckoned in near future. Oil Gas J. 100 (46), 46–48.

Deffeyes, K.S., 2001. Hubbert's Peak—The Impending World Oil Shortage. Princeton University Press, Princeton, NJ, 208 pp.

Delaune, R.D., Baumann, R.H., Gosselink, J.G., 1983. Relationships among vertical accretion, apparent sea level rise and land loss in a Louisiana Gulf Coast marsh. J. Sediment. Petrol. 53, 147–157.

DeLaune, R.D., Pezeshki, S., 2003. The role of soil organic carbon in maintaining surface elevation in rapidly subsiding U.S. Gulf of Mexico coastal marshes. Water Air Soil Pollut. 3, 167–179.

DeLaune, R.D., Jugsujinda, A., Peterson, G., Patrick, W., 2003. Impact of Mississippi River freshwater reintroduction on enhancing marsh accretionary processes in a Louisiana estuary. Estuarine Coastal Shelf Sci. 58, 653–662.

Dingler, J.R., Hsu, S.A., Reiss, T.E., 1992. Theoretical and measured aeolian sand transport on a barrier-island, Louisiana, USA. Sedimentology 39, 1031–1043.

Gornitz, V., Lebedeff, S., Hansen, J., 1982. Global sea level trend in the past century. Science 215, 1611–1614.

Hall, C.A.S., Tharakan, P., Hallock, J., Cleveland, C., Jefferson, M., 2003. Hydrocarbons and the evolution of human culture. Nature 426, 318–322.

Hatton, R.S., DeLaune, R.D., Patrick Jr., W.H., 1983. Sedimentation, accretion and subsidence in marshes of Barataria Basin, Louisiana. Limnol. Oceanogr. 28, 494–502.

Heinberg, R., 2003. The Party's Over—Oil, War and the Fate of Industrial Societies. New Society Publishers, Gabriola Island, Canada, 275 pp.

Hester, M.W., Wilsey, B.J., Mendelssohn, I.A., 1994. Grazing of panicum-amarum in a Louisiana barrier-island dune plant community—management implications for dune restoration projects. Ocean Coastal Manage. 23, 213–224.

IPCC (Intergovernmental Panel on Climate Change), 2001. Climate Change 2001: The Scientific Basis Contribution of Working Group 1 to the Third Assessment Report. Cambridge University Press, Cambridge, UK.

Justic, D., Turner, R.E., Rabalais, N.N., 2003. Climate influences on riverine nitrate flux: implications for coastal marine eutrophication and hypoxia. Estuaries 26, 1–11.

Kerr, R.A., 1998. The next oil crisis looms large-and perhaps close. Science 281, 1128–1131.

Kesel, R.H., 1988. The decline in the suspended load of the lower Mississippi River and its influence on adjacent wetlands. Environ. Geol. Water Sci. 11, 271–281.

Kesel, R.H., 1989. The role of the Mississippi River in wetland loss in Southeastern Louisiana, USA. Environ. Geol. Water Sci. 13, 183–193.

Lane, R.R., Day, J.W., Thibodeaux, B., 1999. Water quality analysis of a freshwater diversion at Caervarvon, Louisiana. Estuaries 22, 327–336.

Lane, R.R., Day, J.W., Marx, B., Reyes, E., Kemp, G.P., 2001. Seasonal and spatial water quality changes in the outflow plume of the Atchafalaya River, Louisiana, USA. Estuaries 25, 30–42.

Lane, R.R., Day, J.W., Justic, D., Reyes, E., Marx, B., Day, J.N., Hyfield, E., 2004. Changes in stoichiometric Si, N and P ratios of Mississippi River water diverted through coastal wetlands to the Gulf of Mexico. Estuarine Coastal Shelf Sci. 60, 1–10.

LA DNR (Louisiana Department of Natural Resources), 1997. The 1997 Evaluation Report to the U.S. Congress on the Effectiveness of Louisiana Coastal Wetland Restoration Projects. Baton Rouge, LA.

LCWCRTF (Louisiana Coastal Wetlands Conservation and Restoration Task Force), 1993. Louisiana Coastal Wetlands Restoration Plan. Main Report. EIS and Appendices. US Army Corps of Engineers, Baton Rouge, LA.

Lindscombe, G., Chabreck, R.H., Hartley, S., 2001. Louisiana Coastal Marsh—Vegetative Type Map. U.S. Geological Survey. National Wetlands Research Center, Lafayette, LA, and the Louisiana Department of Wildlife and Fisheries, Fur and Refuge Division, Baton Rouge, LA (http://www.brownmarsh.net/data/III_9.htm).

Martin, J., Reyes, E., Kemp, G.P., Mashriqui, H., Day, J.W., 2002. Landscape modeling of the Mississippi delta. BioScience 52, 357–365.

Masters, C.D., Root, D.H., Attanasi, E.D., 1991. Resource constraints in petroleum production potential. Science 253, 146–152.

McKee, K., Mendelssohn, I.A., Materne, M.D., 2004. Acute salt marsh dieback in the Mississippi River deltaic plain: a drought induced phenomenon? Global Ecol. Biogeogr. 13, 65–73.

Mendelssohn, I.A., Hester, M.W., Monteferrante, F.J., Talbot, F., 1991. Experimental dune building and vegetative stabilization in a sand-deficient barrier-island setting on the Louisiana coast, USA. J. Coastal Res. 7, 137–149.

Mendelssohn, I.A., Morris, J.T., 2000. Eco-physiological controls on the productivity of Spartina alterniflora loisel. In: Weinstein, M.P., Kreeger, D.A. (Eds.), Concepts and Controversies in Tidal Marsh Ecology. Kluwer Academic Publishers, Boston, MA, pp. 59–80.

Mendelssohn, I.A., Kuhn, N., 2003. Sediment subsidy: effects on soil–plant responses in a rapidly submerging coastal salt marsh. Ecol. Eng. 21, 115–128.

Mitsch, W.J., Day, J.W., Gilliam, J., Groffman, P., Hey, D., Randall, G., Wang, N., 2001. Reducing nitrogen loading to the Gulf of Mexico from the Mississippi River basin: strategies to counter a persistent problem. BioScience 51, 373–388.

Mitsch, W.J., Jørgensen, S.E., 2003. Ecological Engineering and Ecosystem Restoration. Wiley, New York, NY, 411 pp.

Morton, R.A., Buster, N., Krohn, M.D., 2002. Subsurface controls on historical subsidence rates and associated wetland loss in south-central Louisiana. Gulf Coast Assoc. Geol. Soc. 52, 767–778.

Odum, H.T., 1971. Environment, Power and Society. Wiley, New York, NY, 331 pp.

Perez, B.C., Day, J., Justic, D., Twilley, R., 2003. Nitrogen and phosphorus transport between Fourleague bay, Louisiana and the Gulf of Mexico: the role of winter cold fronts and Atchafalaya River discharge. Estuarine Coastal Shelf Sci. 57, 1065–1078.

Poff, L., Brinson, M., Day, J.W., 2002. Aquatic Ecosystems and Global Climate Change: Potential Impacts on Inland Freshwater and Coastal Wetland Ecosystems in the United States. Pew Center on Global Climate Change, Arlington, VA, 45 pp.

Rabalais, N.N., Turner, R.E., Justic, D., Dortch, Q., Wiseman, W.J., Sen Gupta, B.K., 1996. Nutrient changes in the Mississippi River and system responses on the adjacent continental shelf. Estuaries 17, 850–861.

Reed, D., 1992. Effect of weirs on sediment deposition in Louisiana coastal marshes. Environ. Manage. 16, 55–65.

Roberts, H.H., 1997. Dynamic changes of the holocene Mississippi river delta plain: the delta cycle. J. Coastal Res. 13, 605–627.

St. Pé, K., 1999. Interview with the WaterMarks. WaterMarks, p. 11.

Stone, G.W., Grymes, J.M., Dingler, J.R., Pepper, D.A., 1997. Overview and significance of hurricanes on the Louisiana coast, USA. J. Coastal Res. 13, 656–669.

Swenson, E., Turner, R.E., 1987. Spoil banks: effects on a coastal marsh water level regime. Estuarine Coastal Shelf Sci. 24, 599–609.

Thieler, R.R., Hammar-Klose, E.S., 2001. National Assessment of Coastal Vulnerability to Sea-Level Rise: Preliminary Results for

the U.S. Gulf of Mexico Coast. U.S. Geological Survey Open-File Report 00-179 (http://pubs.usgs.gov/of/of00-179).

Turner, R.E., Costanza, R., Scaife, W., 1982. Canals and wetland erosion rates in coastal Louisiana. In: Boesch, D. (Ed.), Proceedings of the Conference on Coastal Erosion and Wetland Modification in Louisiana: Causes, Consequences, and Options. FEW/OBS-82/59. US Fish and Wildlife Service, Office of Biological Services, Slidell, Louisiana, pp. 73–84.

Turner, R.E., Cahoon, D., 1987. Causes of Wetland Loss in the Coastal Central Gulf of Mexico, vol. 1: Executive Summary, vol. 2: Technical Narrative, vol. 3: Appendices. Coastal Ecology Institute, Louisiana State University, Baton Rouge, LA. Final Report Submitted to the Minerals Management Service, Contract No. 14-12-001-30252. OCS Study/MMS 87-0119. New Orleans, LA.

Turner, R.E., Streever, B., 2002. Approaches to Coastal Wetland Restoration: Northern Gulf of Mexico. SPB Academic Publishing, Hague, The Netherlands, 147 pp.

Viosca, P., 1927. Flood control in the Mississippi valley in its relation to Louisiana fisheries. Trans. Am. Fish. Soc. 57, 45–64.

Wolock, D.M., McCabe, G.J., 1999. Estimates of runoff using water-balance and atmospheric general circulation models. J. Am. Water Res. Assoc. 35, 1341–1350.

Available online at www.sciencedirect.com

SCIENCE ***DIRECT***°

ELSEVIER

Ecological Engineering 24 (2005) 267–278

ECOLOGICAL ENGINEERING

www.elsevier.com/locate/ecoleng

Nitrate-nitrogen retention in wetlands in the Mississippi River Basin

William J. Mitsch [a,*], John W. Day [b], Li Zhang [a], Robert R. Lane [b]

[a] *Olentangy River Wetland Research Park, School of Natural Resources, The Ohio State University, 352 W. Dodridge Street, Columbus, OH 43202, USA*

[b] *Department of Oceanography and Coastal Sciences, Coastal Ecology Institute, School of the Coast and Environment, Louisiana State University, Baton Rouge, LA 70803, USA*

Abstract

Nitrate-nitrogen retention as a result of river water diversions is compared in experimental wetland basins in Ohio for 18 wetland-years (9 years × 2 wetland basins) and a large wetland complex in Louisiana (1 wetland basin × 4 years). The Ohio wetlands had an average nitrate-nitrogen retention of $39\,g\text{-}N\,m^{-2}\,year^{-1}$, while the Louisiana wetland had a slightly higher retention of $46\,g\text{-}N\,m^{-2}\,year^{-1}$ for a similar loading rate area. When annual nitrate retention data from these sites are combined with 26 additional wetland-years of data from other wetland sites in the Basin Mississippi River (Ohio, Illinois, and Louisiana), a robust regression model of nitrate retention versus nitrate loading is developed. The model provides an estimate of $22,000\,km^2$ of wetland creation and restoration needed in the Mississippi River Basin to remove 40% of the nitrogen estimated to discharge into the Gulf of Mexico from the river basin. This estimated wetland restoration is 65 times the published net gain of wetlands in the entire USA over the past 10 years as enforced by the Clean Water Act and is four times the cumulative total of the USDA Wetland Reserve Program wetland protection and restoration activity for the entire USA.
© 2005 Published by Elsevier B.V.

Keywords: Wetland; Mississippi River Basin; Restoration; Olentangy River Wetland Research Park

1. Introduction

Humans have increased reactive nitrogen production, much of which becomes biologically available nitrogen, by over an order of magnitude from 1860 to 2000 (15–165 Tg/year), mainly due to fertilizer production, increased use of nitrogen-fixing organisms and fossil fuel combustion (Vitousek et al., 1997; Galloway et al., 2003). Significant amounts of this excess nitrogen are transported as nitrate-nitrogen to rivers and streams, leading to eutrophication and episodic and persistent hypoxia (dissolved oxygen < 2 mg/L) in coastal waters worldwide (NRC, 2000). For example, the Gulf of Mexico hypoxia in North America routinely reaches an extent of $20,000\,km^2$ (Rabalais, 2002; Rabalais et al., 2002; Scavia et al., 2003; Fig. 1). The connection between this hypoxia and the nitrate-nitrogen released from

* Corresponding author. Tel.: +1 614 292 9774; fax: +1 614 292 9773.
  E-mail address: mitsch.1@osu.edu (W.J. Mitsch).

0925-8574/$ – see front matter © 2005 Published by Elsevier B.V.
doi:10.1016/j.ecoleng.2005.02.005

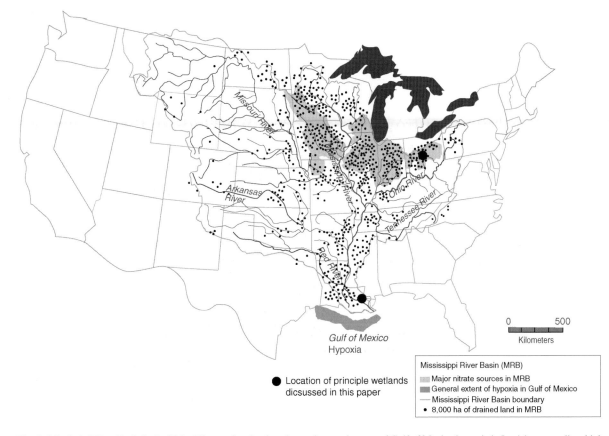

Fig. 1. Mississipi River Basin in the United States, showing location and general extent of Gulf of Mexico hypoxia in Louisiana coastline, high nitrogen loadings in the basin (>1000 kg-N km^{-2} year^{-1}; *source:* Goolsby et al., 1999), major historical drainage in the Basin (*source:* Mitsch and Gosselink, 2000) and wetland sites discussed in this paper (Large black circles).

the 3 million km^2 Mississippi River Basin is well established (Goolsby et al., 1999; McIssac et al., 2002; Rabalais et al., 2002; Scavia et al., 2003). The hypoxia of the Gulf of Mexico is also related to the large loss of wetlands in the Basin (Fig. 1) and to the separation of the Mississippi River from its floodplain and deltaic plain (Dahl, 1990; Mitsch et al., 2001; Day et al., 2003).

Three general approaches for reducing agriculturally derived nitrogen that would otherwise reach the Gulf of Mexico are (Mitsch et al., 2001): (1) change farming practices to minimize nitrate loss by reducing the use of nitrogen fertilizer and through a suite of

management practices, (2) intercept laterally moving groundwater and surface water with nitrogen-sink ecosystems, particularly riparian zones and created and restored wetlands and (3) provide a system of river diversion backwaters along rivers and in the Mississippi River delta for interception of large fluxes of nitrogen associated with flood events. This paper concerns the efficacy of the second and third options and presents long-term data records that establish similarities in function of wetland retention of nitrate-nitrogen at different scales and climates in the Mississippi River Basin. Wetlands and riparian ecosystems can serve

Fig. 2. Two wetland research locations discussed in detail in this paper: (a) two 1-ha experimental wetlands at Olentangy River Wetland Research Park, Columbus, Ohio and (b) Caernarvon diversion to Breton Sound at Mississippi River in Louisiana (shading indicates area of most significant influence of diversion). Sampling locations in (a) were at the inflow to experimental wetland 1 (since the same water was delivered to each wetland) and at the outflows of wetlands 1 and 2. Sampling stations in (b) were at various locations southeast of the diversion and toward Breton Sound.

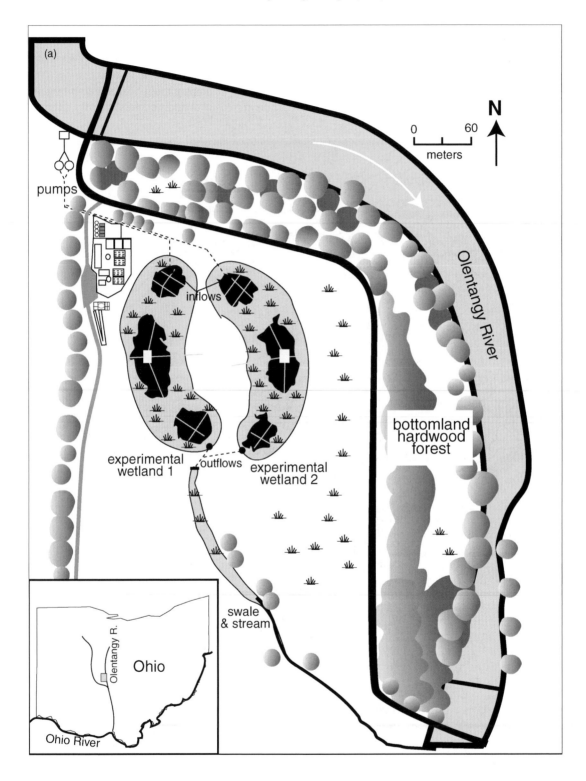

as buffers between agricultural uplands and streams and rivers, particularly for excessive nitrate-nitrogen emanating from fertilizer use (Mitsch et al., 2001; Day et al., 2003). They can be designed in the landscape to enhance nitrate-nitrogen reduction through two main ecological processes: (1) denitrification and (2)

nitrogen uptake by plants, microbes and macrophytes. The latter process is important if nitrogen is subsequently buried in the soil or if the plant material is permanently retained or harvested.

We first compare multi-year nitrate-nitrogen retention in river diversion wetlands at vastly different river

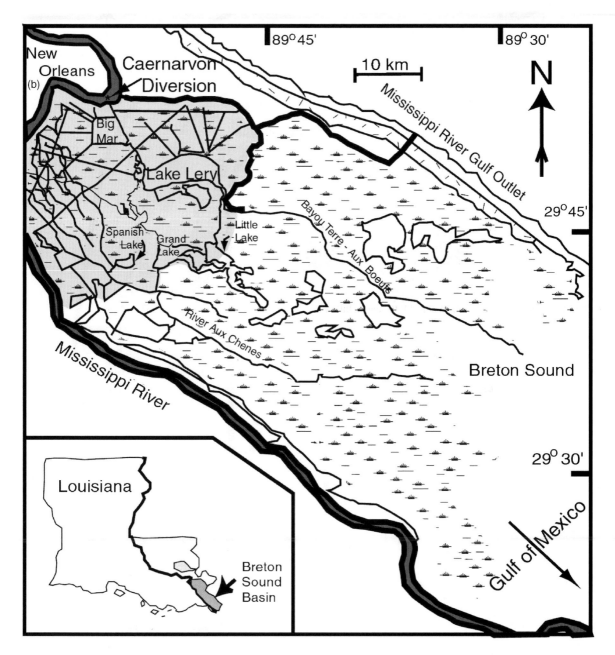

Fig. 2. (*Continued*).

scales—a fourth-order river in Ohio and the ninth-order lower Mississippi River at its delta in Louisiana. A river diversion wetland is a wetland on the adjacent floodplain or behind artificial levees that receives water by pumping or gravity flow from the main channel of a river and includes such floodplain features as oxbow lakes, backwater swamps and other riparian wetlands. Both the Ohio and Louisiana sites are part of the Mississippi River Basin that drains to the Gulf of Mexico (Fig. 1). Each riparian diversion system has been the site of significant research with similar sampling and analytical methodologies for several years. Each project involves diverting nutrient-laden riverine waters into adjacent riparian wetlands.

2. Materials and methods

2.1. Study areas

In Ohio, a pair of 1-ha experimental diversion wetlands basins were created in 1993 (Fig. 2a) and used in a whole-ecosystem wetland experiment from 1994 to 2003 (Mitsch et al., 1998, 2005; Mitsch and Jørgensen, 2004). Continuously pumped inflows have averaged 0.006–0.010 $m^3 s^{-1}$ (20–30 m year^{-1}) to each basin with day-to-day flow patterns corresponding to Olentangy River flow. The water passes through the wetlands in about 3–4 days and discharges to a common swale that in turn flows back to the Olentangy River.

In Louisiana, the diversion of the Mississippi River at Caernarvon (Fig. 2b) is one of the largest diversions in operation on the River aimed at restoring deteriorating wetlands in the Mississippi delta. The diversion structure on the east bank of the river south of New Orleans is a five-box culvert with vertical lift gates with a maximum flow of 280 $m^3 s^{-1}$. River diversion began in August 1991 and average minimum and maximum flows are 14 and 114 $m^3 s^{-1}$, respectively, with summer flow rates generally near the minimum and winter flow rates 50–80% of the maximum (Lane et al., 1999, 2004). The diversion delivers river water to the 260 km^2 Caernarvon freshwater wetland that eventually discharges into the brackish Breton Sound estuary, which is a part of coastal Gulf of Mexico.

In both cases, significant infrastructure (diversion gates, retention valves, plumbing and pumps) were used to control and measure flows into adjacent riparian wetlands. In both diversion systems, the wetlands are dominated by marshes, the percent macrophyte cover is similar at about 60% and net primary productivity at both sites is comparable.

2.2. Sampling and analysis

Weekly water samples were taken from the Olentangy River near the inflow pumps and at the inflow and outflows of the two experimental wetlands in Ohio for 9 years from 1994 to 2001 and 2003 (Fig. 1). Monthly water samples were taken along the major flow paths in the Caernarvon wetland in Louisiana from 1991 to 1994. Mississippi River nitrate data were collected and analyzed from 1988 to 1994 (7 years). During spring of 2001, an experimental large pulse of river water with a peak flow of 226 $m^3 s^{-1}$ was released through the Caernarvon structure for 16 days. Sampling was carried out during weekly transects from March 9 to 30, 2001. Discrete water samples were taken at 19 locations in the Breton Sound estuary, but only the five sampling locations closest to the diversion structure were used in this analysis (Lane et al., 2004).

Sample analysis at both sites was carried out using standard analytical methods (U.S. EPA, 1983, APHA, 1989, 1992). Water samples were collected in acid-washed bottles, filtered through 0.45 μm filters and frozen for later analysis of nitrate + nitrite ($NO_3 + NO_2$). Nitrate + nitrite were analyzed on a Lachat QuikChem IV automated system in Ohio and on a Alpkem autoanalyzer in Louisiana using the cadmium reduction method. Samples from April 1994 through July 1995 were run by similar methods by Heidelberg College Water Quality Laboratory using a Traacs 800 autoanalyzer. The accuracy of the nutrient analysis was checked every 10–20 samples with a known standard and the samples are redone if the accuracy was off by 5%.

Twice-daily (morning and evening) readings of both instantaneous and total integrated volume of pumping rates were collected by staff and students from the flow monitors in each pipe going to each Ohio wetland. Outflow measurements from the experimental wetlands are based on wetland water level and the status of the control weir boxes constructed at the southern edge of the basins. Manual readings of water level data were supplemented with continuous water

level Ott Thalimedes data loggers installed in 2001 in each Ohio wetland basin.

2.3. Nutrient loading calculations

Nutrient loading rate (expressed as g-N m^{-2} year^{-1}) and removal efficiency (the percentage of nutrients removed from the water column based on both concentration and mass) were calculated for each wetland site in a similar manner.

Nutrient retention of nitrate was calculated using,

Nutrient reduction (% by mass)

$$= \left(\frac{Q_{in} - Q_{out}}{Q_{in}} \right) \times 100 \qquad (1)$$

where Q_{in} is the inflow flux of nitrate-nitrogen in the incoming river water and Q_{out} is the outflow flux of nitrate-nitrogen from the wetlands.

In Louisiana, since the wetland does not have a formal "basin" but rather extends eventually into Breton Sound and the Gulf of Mexico, the portion of wetlands between the Caernarvon Diversion structure and several water quality stations was used for calculating nutrient reduction (Lane et al., 1999). The effective area of wetlands was estimated to be 260 km^2. Two different datasets were used to analyze nitrate loading and

retention at the Caernarvon diversion area: a monthly dataset from 1992 to 1994 (also analyzed by Lane et al., 1999) and a 1-year dataset with 15 sampling dates taken in 2001. Discharge from the Caernarvon diversion and NO$_3$ concentrations of incoming Mississippi River water were used to calculate nitrate-nitrogen inflow into the wetland. Two end-member stations were used in the 1992–1994 analysis and five were used with the 2001 dataset. Since water flows through two major routes (referred to as eastern and western), with approximately 66% of the flow being carried down the eastern route, the data were weighted accordingly to reflect the proportion of flow each route conveyed. The wetland areas used to calculate nitrate-nitrogen fluxes were 260 km^2 for the 1992–1994 data and 10, 30 and 50 km^2 for the 2001 data. Different areas were used in 2001 because of more detailed sampling at more sampling stations.

2.4. Data analysis

Statistical analyses were computed by SPSS 11.0 software, e.g. predictions for reduction% of NO$_3$ by concentration and mass versus inflow loading. A regression of curve estimation with logarithmic functions was used to generate predictions for % nitrate reduction with a 95% confidence interval.

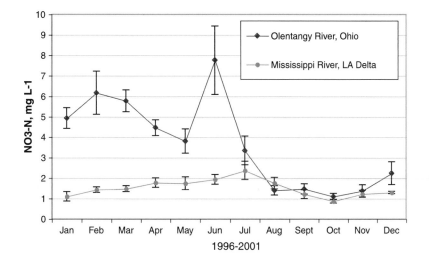

Fig. 3. Nitrate-nitrogen concentrations (average ± S.E.) in Olentangy River, Ohio and Mississippi River, Louisiana. Data for Olentangy River in Columbus are based on 8 years of weekly sampling; data from Mississippi River are based on 7 years of monthly sampling at Caernarvon, Louisiana.

3. Results and discussion

3.1. Patterns of nitrate-nitrogen in the rivers

Nitrate-nitrogen patterns in the Olentangy River in Ohio and Mississippi River in Louisiana are different in the "wet season" of January through June and similar in the "dry season" of late summer and autumn (Fig. 3). The Olentangy River, which is fed by an agricultural and urban watershed, has NO_3-N concentrations of 4–6 mg-N L^{-1} in the spring. The multi-year Olentangy River data reflect the general pattern of nitrate-nitrogen in Midwestern rivers, when peak concentrations are usually coincident with the first sustained precipitation events after fertilizer is applied in the spring (Randall et al., 1997; Goolsby et al., 1999; Mitsch et al., 2001). A peak with a large variability occurs in June when 63% of the weekly measurements over 6 years were greater than 5 mg-N L^{-1}. Nitrate concentrations in the lower Mississippi River near the Caernarvon diversion rise from a low of 1 mg-N L^{-1} in late fall and early winter to about 2 mg-N L^{-1} during high-flow conditions in late spring. Concentrations of nitrate-nitrogen in the Olentangy and Mississippi Rivers are remarkably similar from August through November.

3.2. Comparison of Ohio and Louisiana wetland nitrate retention

To compare the two-wetland sites given their different flow rates and vastly different sizes, inflows were normalized for the size of the wetland, i.e., areal loading rates and retention rates were calculated (Table 1). For 18-wetland-years of measurements (2 wetlands × 9 years), the Ohio wetlands retained an average of 35 ± 2% of nitrate-nitrogen by concentration

Table 1

Nitrate-nitrogen inflow, outflow and retention (by mass and concentration) for Olentangy River diversion wetlands in Ohio and Carenarvon River diversion wetlands in Louisiana

Wetland	Inflow g-N m^{-2} year^{-1}	Outflow g-N m^{-2} year^{-1}	Retention g-N m^{-2} year^{-1}	Mass retention (%)	Concentration retention (%)
Olentangy River wetlands, Ohio					
1994 Wetland 1	57.2	41.6	15.7	27	49
1994 Wetland 2	57.9	45.2	12.7	22	46
1995 Wetland 1	85.8	67.9	17.8	21	36
1995 Wetland 2	85.8	59.9	25.9	30	42
1996 Wetland 1	58.4	39.1	19.3	33	33
1996 Wetland 2	58.5	43.5	15.0	26	25
1997 Wetland 1	211	130	81	38	17
1997 Wetland 2	215	124	91	42	18
1998 Wetland 1	136	95	41	30	33
1998 Wetland 2	138	83	55	40	39
1999 Wetland 1	78.6	57.3	21.3	27	30
1999 Wetland 2	81.9	51.0	30.9	38	33
2000 Wetland 1	129.3	81.2	48.1	37	34
2000 Wetland 2	128.4	80.0	48.4	38	44
2001 Wetland 1	112.2	63.1	49.1	44	35
2001 Wetland 2	106.2	68.9	37.3	35	23
2003 Wetland 1	104.9	62.3	42.6	41	41
2003 Wetland 2	98.7	47.5	51.2	52	44
Caernarvon diversion, Louisiana					
1992[a]	5.60	0.17	5.43	97	97
1993[a]	7.30	1.54	5.76	79	79
1994[a]	12.7	4.2	8.5	67	67
2001[b]	50	19	31	62	62
2001[b]	84	38	46	55	55
2001[b]	251	161	90	36	36

[a] Based on effective Caernarvon wetland area of 226 m^2.

[b] Calculated on basis of Caernarvon wetland area of 10, 30 and 50 km^2 for the 2001 data.

and $35 \pm 2\%$ by mass. Our records showed that, overall, there a variability in concentrations of nitrate-nitrogen in the Olentangy River. In years of relatively high nitrate concentration (1996, 1997 and 2000), nitrates decreased from approximately $4.6–4.9 \, \text{mg-N} \, \text{L}^{-1}$ to $2.4–3.5 \, \text{mg-N} \, \text{L}^{-1}$ in the 1-ha wetlands. In a low-nitrate concentration year (e.g., 1999), nitrates decreased from approximately 2 to $1.3 \, \text{mg-N} \, \text{L}^{-1}$.

By contrast, the Caernarvon wetland retained 39–92% of nitrate by mass and concentration,

depending on the sampling location (Table 1). At the Caernarvon Louisiana sampling station that was most comparable to the Ohio wetlands for loading rates (station at $30 \, \text{km}^2$, where loading rate is $84 \, \text{g-N} \, \text{m}^{-2} \, \text{year}^{-1}$ compared to an average loading rate at the Ohio wetlands of $108 \pm 11 \, \text{g-N} \, \text{m}^{-2} \, \text{year}^{-1}$), the nitrate-nitrogen retention was 55% by mass and concentration (Table 1). The Olentangy River wetland site in Ohio averaged 35% retention by mass and concentration, considerably less retention. The Ohio wetlands

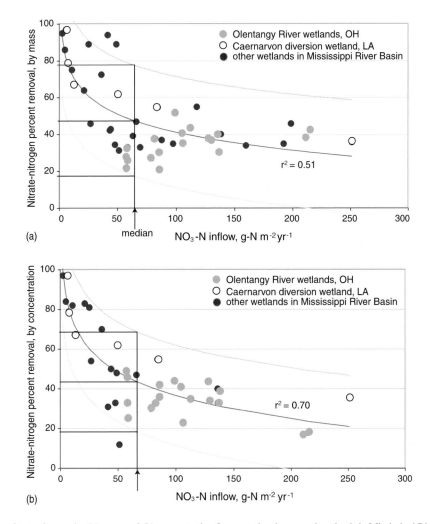

Fig. 4. Decrease in nitrate-nitrogen by (a) mass and (b) concentration for created and managed wetlands in Mississippi River Basin. Each data point represents data for a complete year for a wetland except for Caernarvon 2001data, which are based on 1-year of data (2001) and different "wetland basin" areas. Data from Olentangy River wetlands in Ohio and Caernarvon wetlands in Louisiana are supplemented by data from other wetlands studies in Ohio (Fink and Mitsch, 2004), Illinois (Kovacic et al., 2000; Phipps and Crumpton, 1994) and Louisiana (Lane et al., 2002). Outside lines are 95% confidence intervals. Vertical lines in graphs (a) and (b) indicate median loading rate of $60 \, \text{g-N} \, \text{m}^{-2} \, \text{year}^{-1}$ and three resulting horizontal lines indicate median removal bracketed by the 95% confidence intervals used for predicting required area of wetlands.

had an average mass retention of 39 g-N m^{-2} year^{-1}, while the Louisiana wetland had a slightly higher retention of 46 g-N m^{-2} year^{-1} for a similar loading rate area. We believe that the more subtropical climate in southern Louisiana compared to the continental temperate climate in central Ohio is conducive to higher rates of denitrification and nutrient uptake because of higher water temperatures and a longer growing season.

3.3. Nitrogen retention model

A model of nitrate-nitrogen retention by wetlands in the Mississippi River Basin was developed combining the above 24 wetland-years of nitrate-nitrogen data from Ohio and Louisiana with nitrate reduction data from another 26 wetland-years of data from additional wetlands in Ohio, Illinois and Louisiana (Phipps and Crumpton, 1994; Kovacic et al., 2000; Lane et al., 2002; Fink and Mitsch, 2004; Mitsch and Day, in press). The nonlinear regression model explains 51% of the variation between nitrate-nitrogen inflow per unit area and nitrate mass reduction in the wetlands (Fig. 4a) and 70% of the variation for predicting nitrate concentration reduction in wetlands (Fig. 4b). In all of these additional case studies, the wetlands received either river diversions or agricultural runoff and all had published annual data on retention. Data from more extensively studied wastewater treatment wetlands (e.g., Kadlec and Knight, 1996) were not included here as these wetlands usually involve much higher loading rates and concentrations of nitrogen and can have different nitrogen species dominating (e.g., ammonium-nitrogen rather than nitrate-nitrogen) than normally found in ambient river and agricultural runoff.

3.4. The scale of wetland creation and restoration needed

Based on the relationship between loading and retention rates for wetlands, we determined from 50 wetland-years of data from 12 independent wetland basins in the Mississippi River Basin in Fig. 4, we estimated the wetland area required to remove 40% of the nitrogen load to the Gulf of Mexico. We assumed a median inflow rate of 60 g-N m^{-2} year^{-1} (and hence from Fig. 4a a nitrate-nirogen retention rate of 48% or 29 g-N m^{-2} year^{-1}). Creation or restoration of 22,000 km^2 of wetlands in the Mississippi River Basin is estimated as the area required to remove 40% of

the total nitrogen estimated by Goolsby et al. (1999) calculated as discharging to the Gulf of Mexico (total nitrogen flux to the Gulf of Mexico = 1,570,000 Mg-N year^{-1}). The calculation is shown as:

$$\text{Area required (km}^2)$$
$$= \frac{1.57 \times 10^{12}(\text{g-N year}^{-1}) \times 0.40}{29(\text{g-N m}^{-2}\,\text{year}^{-1}) \times 10^6(\text{m}^2\,\text{km}^{-2})}$$
$$= 22,000\,\text{km}^2$$

In order to attach a variance on this estimate, we developed a model that predicted the 95% confidence interval for the data as shown in Fig. 4a. Using that confidence interval with the same loading assumption, the amount of wetlands needed to reduce the load by 40% ranges from 13,000 to 58,000 km^2. The low area assumes a mass retention of 78% of nitrate-nitrogen in the wetlands and the high area assumes a mass retention of 18% of nitrate-nitrogen in the wetlands as shown in Fig. 4a.

Assuming the same inflow rate of nitrate-nitrogen as described above, our model predicts that the average wetland might be expected to reduce nitrate-nitrogen concentrations about 45%, with a 95% confidence interval between 19 and 68% reduction (Fig. 4b). Nitrate-nitrogen retention rates above 80% should not be expected unless inflow rates are a third of our design inflow rate of 60 g-N m^{-2} year^{-1}.

If bottomland hardwood riparian forests are used as "wetlands" in this management strategy, the area required for an equivalent nitrate-nitrogen reduction is probably more, as our analysis has shown that bottomland hardwood forests retain generally less nitrogen per unit area than do wetlands (Mitsch et al., 2001). Our 22,000 km^2 estimate is based on more complete wetland data than were available in the late 1990s and provides a more confident estimate of the required wetland area than an earlier estimate of 21,000–53,000 km^2 of wetlands recommended by us earlier (Mitsch et al., 2001). This ecologically engineered nitrogen reduction, combined with an estimated 20% nitrate reduction that we estimated could be done by appropriate agronomic practices would result in enough reduction in nitrates entering the Gulf of Mexico to ensure a significant decrease in the size of the Gulf of Mexico hypoxia. We base this conclusion

on the fact that Brezonik et al. (1999) found a generally linear relationship between nitrate-nitrogen flux to the Gulf and the size of the hypoxia in the Gulf. An overall reduction of 60% of nitrate-nitrogen in the Mississippi River Basin due to agronomic and ecological means should reduce the hypoxia area by approximately 60%.

3.5. Comparing this restoration to other wetland restoration efforts

To put our wetland restoration recommendation of 22,000 km^2 in perspective, there has been an estimated net gain of only 336 km^2 of wetlands in the entire United States over the past decade (*source:* US ACOE 2003 data, personal communication) due to wetland mitigation and enforcement of the Clean Water Act. By contrast, under a USDA national conservation set-aside program in agriculture to restore and protect wetlands called the Wetland Reserve Program (WRP), it is estimated that farmers have restored about 6000 km^2 of wetlands in the United States though 2003 (*source:* Natural Resources Conservation Service 2004 web site: http://www.nrcs.usda.gov/programs/wrp/). Thus, an effort estimated to be 65 times current efforts of wetland restoration and creation in the entire United States through the Clean Water Act and 4 times current wetland restoration through the Wetland Reserve Program would be needed in the Mississippi River Basin alone to have a significant effect on the Gulf of Mexico hypoxia.

3.6. Wetlands and agriculture

Our recommendation of setting aside less than 1% or 22,000 km^2 of the 3,000,000 km^2 Mississippi River Basin as an ecological solution to the Gulf of Mexico hypoxia provides a reasonable alternative to a major reduction in fertilizer use that could reduce agricultural output and cause an economic impact in the basin. Doering et al. (1999) agreed with this assessment when they found that restoring 20,000 km^2 of wetlands in the Mississippi River Basin would have minimal impact on agricultural production in the basin. Furthermore, they argued that if nitrogen fertilizer use is restricted within the Mississippi River Basin, then the crop production and hence nitrogen pollution would simply be transferred to somewhere else. Our solution to the nitrate problem does not lead to the transfer of nitrate pollution to another watershed.

McIssac et al. (2002) suggested by retrospective analysis that a 33% reduction in nitrate-nitrogen in the basin could result from simply reducing nitrogen fertilizer use in the basin by 12%. We believe that this is a misleading interpretation that makes solving the problem appear to be much easier than it really is. There are uncertainties in the data as described in Goolsby et al. (1999) that were used in this analysis. Studies described by Randall et al. (1997) and Mitsch et al. (1999) in Minnesota and R. Turco (personal communication) in Indiana show a relative insensitivity of nitrate-nitrogen in the effluent of test agronomic plots to fertilizer use rates. For example, plot data in Minnesota showed that, "if the annual nitrogen fertilizer rate was reduced by about 10% to 125 kg-N/ha and no other nitrogen was applied, one could expect a small yield decrease and nitrate concentrations could be expected to decrease about 3 mg-N/L" (Mitsch et al., 1999). This was a relatively low (~10%) decrease in nitrate-nitrogen in the effluent from these study plots. Both agronomic and ecological solutions are needed together for this large-scale pollution problem.

3.7. Benefits to the basin

The wetland creation and restoration suggested here is primarily for solving the problem of the Gulf of Mexico hypoxia off of the coast of Louisiana. But the wetland restoration recommended in our paper would also provide other ecological services locally in the upper Mississippi River Basin including restoration of wetland and riverine habitats and provision of flood mitigation, both of which are very much needed in the Mississippi River Basin because of past wetland drainage in the Basin (Mitsch and Day, in press). The nitrate-nitrogen reduction in Midwestern rivers caused by both wetlands and agronomic practices would also lessen public health concerns about nitrate-nitrogen in urban drinking water taken from Midwestern rivers.

Acknowledgements

The work was supported by grants from Louisiana Department of Natural Resources (Contract No. 2512-03-11), Ohio Agricultural Research and Development Center Payne Grant No. 2002-079, U.S. Department of Agriculture (Nutrient Science Program Grant No.

2002-51130-1919 and National Research Initiative CSREES Award 2003-35102-13518) and U.S. Environmental Protection Agency Water and Watersheds Grant R-82900901-0. Some support was also provided by the U.S. Army Corps of Engineers. We appreciate the useful comments provided by several reviewers and the review and updated data on some of the central Illinois wetlands provided by David Kobacic. This paper is Olentangy River Wetland Research Park Publication 05-004.

References

American Public Health Association, 1989. Standard Methods for the Analysis of Wastewater, 17th ed. APHA, Washington, DC.

American Public Health Association, 1992. Standard Methods for the Analysis of Wastewater, 18th ed. APHA, Washington, DC.

Brezonik, P.L., Bierman, V.J., Alexander, R., Anderson, J., Barko, J., Dortch, M., Hatch, L., Hitchcock, G.L., Keeney, D., Mulla, D., Smith, V., Walker, C., Whitledge, T., Wiseman, W.J., 1999. Effects of reducing nutrient loads to surface waters within the Mississippi River Basin and the Gulf of Mexico. Topic 4 Report for the Integrated Assessment of Hypoxia in the Gulf of Mexico. NOAA Coastal Ocean Program Decision Analysis Series No. 18. NOAA Coastal Ocean Program, Silver Spring, MD, 130 pp.

Dahl, T.E., 1980. Wetland Losses in the United States 1780s to 1980s. U.S. Department of Interior, Fish and Wildlife Service, Washington, DC, 21 pp.

Day Jr., J.W., Yañéz Arancibia, A., Mitsch, W.J., Lara-Dominguez, A.L., Day, J.N., Ko, J.-Y., Lane, R., Lindsey, J., 2003. Using ecotechnology to address water quality and wetland habitat loss problems in the Mississippi Basin: a hierarchical approach. Biotechnol. Adv. 22, 135–159.

Doering, O.C., Diaz-Hermelo, F., Howard, C., Heimlich, R., Hitzhusen, F., Kazmierczak, R., Lee, J., Libby, L., Milon, W., Prato, T., Ribaudo, M., 1999. Evaluation of the economic costs and benefits of methods for reducing nutrient loads to the Gulf of Mexico. Topic 6 Report for the Integrated Assessment of Hypoxia in the Gulf of Mexico. NOAA Coastal Ocean Program Decision Analysis Series No. 20. NOAA Coastal Ocean Program, Silver Spring, MD, 115 pp.

Fink, D.F., Mitsch, W.J., 2004. Seasonal and storm event nutrient removal by a created wetland in an agricultural watershed. Ecol. Eng. 23, 313–325.

Galloway, J.N., Aber, J.D., Erisman, J.W., Seitzinger, S.P., Howarth, R.W., Cowling, E.B., Cosby, B.J., 2003. The nitrogen cascade. BioScience 53, 341–356.

Goolsby, D.A., Battaglin, W.A., Lawrence, G.B., Artz, R.S., Aulenbach, B.T., Hooper, R.P., Keeney, D.R., Stensland, G.I., 1999. Flux and sources of nutrients in the Mississippi-Atchafalaya River Basin. Topic 3 Report for the Integrated Assessment of Hypoxia in the Gulf of Mexico. National Oceanographic and Atmospheric Administration Coastal Ocean Program Decision Analysis Series No. 17, NOAA, Silver Springs, MD.

Kadlec, R., Knight, R., 1996. Treatment Wetlands. CRC Press, Boca Raton, FL.

Kovacic, D.A., David, M.B., Gentry, L.E., Starks, K.M., Cooke, R.A., 2000. Effectiveness of constructed wetlands in reducing nitrogen and phosphorus export from agricultural tile drainage. J. Environ. Qual. 29, 1262–1274.

Lane, R.R., Day, J.W., Thibodeaux, B., 1999. Water quality analysis of a freshwater diversion at Caernarvon, Louisiana. Estuaries 22, 327–336.

Lane, R.R., Day, J.W., Kemp, G., Marx, B., 2002. Seasonal and spatial water quality changes in the outflow plume of the Atchafalaya River, Louisiana. Estuaries 25, 30–42.

Lane, R., Day, J.W., Justic, D., Reyes, E., Marx, B., Day, J.N., Hyfield, E., 2004. Changes in stoichiometric Si, N, and P ratios of Mississippi River water diverting through coastal wetlands to the Gulf of Mexico. Estuarine, Coastal Shelf Sci. 60, 1–10.

McIssac, G.F., David, M.B., Gertner, G.Z., Goolsby, D.A., 2002. Nitrate flux in the Mississippi River. Nature 414, 166–167.

Mitsch, W.J., Wu, X., Nairn, R.W., Weihe, P.E., Wang, N., Deal, R., Boucher, C.E., 1998. Creating and restoring wetlands: a whole-ecosystem experiment in self-design. BioScience 48, 1019–1030.

Mitsch, W.J., Day, J.W., Gilliam, J.W., Jr., Groffman, P.M., Hey, D.L., Randall, G.W., Wang, N., 1999. Reducing nutrient loads, especially nitrate-nitrogen, to surface water, groundwater, and the Gulf of Mexico. Topic 5 Report for the Integrated Assessment on Hypoxia in the Gulf of Mexico. NOAA Coastal Ocean Program Decision Analysis Series No. 19. NOAA Coastal Ocean Program, Silver Spring, MD, 111 pp.

Mitsch, W.J., Day Jr., J.W., Gilliam, J.W., Groffman, P.M., Hey, D.L., Randall, G.W., Wang, N., 2001. Reducing nitrogen loading to the Gulf of Mexico from the Mississippi River Basin: strategies to counter a persistent ecological problem. BioScience 51, 373–388.

Mitsch, W.J., Gosselink, J.G., 2000. Wetlands, third ed. John Wiley & Sons, New York, 920 pp.

Mitsch, W.J., Jørgensen, S.E., 2004. Ecological Engineering and Ecosystem Restoration. John Wiley & Sons, New York, 411 pp.

Mitsch, W.J., Wang, N., Zhang, L., Deal, R., Wu, X., Zuwerink, A., 2005. Using ecological indicators in a whole-ecosystem wetland experiment. In: Jørgensen, S.E., Xu, F.-L., Costanza, R. (Eds.), Handbook of Ecological Indicators for Assessment of Ecosystem Health. CRC Press, Boca Raton, FL, pp. 211–235.

Mitsch, W.J., Day, J.W., Jr. Ecological restoration of the Mississippi-Ohio-Missouri River Basin. Ecol. Eng., in press.

National Research Council, 2000. Clean Coastal Waters: Understanding and Reducing the Effects of Nutrient Pollution. National Academy Press, Washington, DC.

Phipps, R.G., Crumpton, W.G., 1994. Factors affecting nitrogen loss in experimental wetlands with different hydrologic loads. Ecol. Eng. 3, 399–408.

Rabalais, N.N., 2002. Nitrogen in aquatic ecosystems. Ambio 31, 102–112.

Rabalais, N.N., Turner, R.E., Scavia, D., 2002. Beyond science into policy: Gulf of Mexico hypoxia and the Mississippi River. BioScience 52, 129–142.

Randall, G.W., Huggins, D.R., Russelle, M.P., Fuchs, D.J., Nelson, W.W., Anderson, J.L., 1997. Nitrate losses through subsurface tile drainage in CRP, alfalfa and row crop systems. J. Environ. Qual. 26, 1240–1247.

Scavia, D., Rabalais, N.N., Turner, R.E., Justic, D., Wiseman, W.J., 2003. Predicting the response of Gulf of Mexico hypoxia to variations in Mississippi River load. Limnol. Oceanogr. 48, 951–956.

U.S. Environmental Protection Agency, 1983. Handbook for Methods in Water and Wastewater Analysis. U.S. Environmental Protection Agency, Cincinnati, OH.

Vitousek, P.M., Howarth, R.W., Likens, G.E., Matson, P.A., Schindler, D., Schlesinger, W.H., Tilman, G.D., 1997. Human alteration of the global nitrogen cycle: causes and consequences. Issues Ecol. 1, 1–17.

Available online at www.sciencedirect.com

SCIENCE  DIRECT°

ELSEVIER

Ecological Engineering 24 (2005) 279–287

ECOLOGICAL ENGINEERING

www.elsevier.com/locate/ecoleng

Nutrient farming: The business of environmental management

Donald L. Hey*, Laura S. Urban, Jill A. Kostel

Senior Vice President, The Wetlands Initiative, 53 W. Jackson Blvd., #1015, Chicago, IL 60604, USA

Received 29 April 2004; received in revised form 23 September 2004; accepted 1 November 2004

Abstract

Restored wetlands could be used successfully to address our recurring problems of excess nutrients (and sediments) and flood damages along U.S. rivers. Credit markets for flood storage, nitrogen, phosphorous, carbon, atrazine, sediment, and many other constituents would economically motivate landowners to restore wetlands. The resulting high-quality open space would provide for recreation, wildlife habitat, and biodiversity. By instigating the market for nitrate-nitrogen, we can jumpstart the entire process of using markets to manage ecosystems. The nitrogen market will create a new land-economics paradigm and new opportunities for landowners, particularly farmers.
© 2005 Elsevier B.V. All rights reserved.

Keywords: Nutrient control; Flood control; Restored wetlands; Credit market

1. Introduction

Our nation can address and ameliorate several major ecosystem problems (e.g., flooding, excess nutrients, habitat loss) by restoring wetlands. Restored wetlands can store floodwaters, remove excess nutrients, and provide wildlife habitat. Furthermore, we can finance this restoration by creating nutrient removal credits and selling or trading them to dischargers who need to meet water quality standards. These credits, bought or sold on an open market or through long-term contracts, could offset wetland losses due to agricultural, industrial, commercial, or residential development, while providing high-quality open space

for recreation, wildlife habitat and biodiversity. The Wetlands Initiative calls this strategy of creating a marketplace for nutrient credits "Nutrient Farming." In comparison to other credit-based programs that focus on watershed trading opportunities between municipalities (Moore et al., 2000) or point and non-point sources (Johnson et al., 2001), nutrient farming centers on the use of wetlands to remove excess nutrients.

One of the easiest ecosystem commodities to generate, monitor, and manage is aqueous nitrate–nitrogen (NO_3–N). Wetlands do not sequester nitrate; they remove it largely through denitrification (Mitsch and Gosselink, 2000). An anaerobic biological process, denitrification reduces inorganic nitrate (NO_3^-) and nitrite (NO_2^-) to nitrogen gas (N_2). The biological process is dependent on the microbial communities present in the wetlands. Gaseous nitrogen volatilizes, thus ni-

---

* Corresponding author.
E-mail address: dhey@wetlands-initiative.org (D.L. Hey).

trogen is eliminated as a water pollutant. Nitrogen gas is relatively inert, so its release to the atmosphere poses no danger; in fact, nitrogen comprises 78% of the atmosphere. To a lesser extent, nitrate is also removed from water by assimilative nitrate reduction, where nitrate is reduced to ammonia, which serves as a nitrogen source for growth by plants, fungi, and bacteria.

A viable nitrogen credit market will readily demonstrate the economic value of land for purposes other than corn production, housing, or commercial development. The nitrogen market will monetize ecosystem services for water quality management. This precedent will facilitate monetizing phosphorous control, flood storage, wildlife management, and many other ecosystem services. The nitrogen market will create a new land-economics paradigm. It will create new financial opportunities for landowners, particularly farmers, and lessen the agricultural community's dependence on government subsidies, while bringing an equitable resolution to the problem of non-point source pollution.

2. Excess nutrients

The modern landscape (e.g., the Upper Mississippi River Basin) suffers from degraded water quality, excessive flood damage, decimated wildlife populations, and declining biodiversity. The impact of landscape modifications in this basin also directly affects other regions. For example, the hypoxic zone in the northern Gulf of Mexico has nearly doubled in size in the past two decades. The hypoxic area averaged $8300 \, km^2$ in 1985–1992 and increased to approximately $16,000 \, km^2$ in 1993–2001 (Rabalais et al., 2002). Explanations for the increased size of the hypoxic zone have varied, but the increase is principally attributed to an almost three-fold increase in nitrogen load to the Upper Mississippi River Basin since 1950 (Goolsby et al., 1999). Hypoxia, or oxygen depletion, is the result of the overenrichment of nitrogen, mainly nitrate–nitrogen, which increases algal production within this aquatic ecosystem. The decomposition of the dead algae leads to the low dissolved oxygen concentrations (<2 mg/L O_2). During the late spring and early summer months, the low dissolved oxygen concentrations force fish and other mobile aquatic organisms to flee the regions of low oxygen, while less mobile organisms are killed. High nitrogen concentrations have been linked to

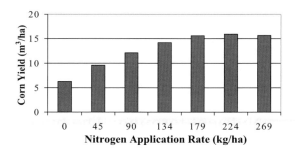

Fig. 1. Effect on corn yield of nitrogen application rate (Hoeft et al., 1999).

human and ecological health effects, including blue baby syndrome and increased risks of cancer (Weyer et al., 2001). In addition to the excess nitrogen loading, wetland loss and more efficient drainage practices have contributed to the hypoxia problem.

Agricultural practices are the principal sources of nitrogen to the basin as the use of commercial fertilizers, the application of manure, and the production of legumes (e.g., soybeans) contribute to the increased nutrient concentrations (Goolsby et al., 1999). In the Mississippi basin, 31% of the nitrogen load to the Gulf of Mexico comes from the fertilizer applied to agricultural lands. Corn yields have increased as nitrogen fertilizer application rates have increased (Fig. 1). Hoeft et al. (1999) found a 2.5 increase in corn yield with application of 224 kg/ha (200 lbs/acre) of nitrogen fertilizer compared to crops with no fertilizer. This increase in yield is not merely theoretical; annual use of nitrogen fertilizer has increased six-fold since 1950 in the Mississippi–Atchafalya River Basin (Fig. 2).

While use of nitrogen-based fertilizers in the basin has increased, so has hydraulic efficiency of the

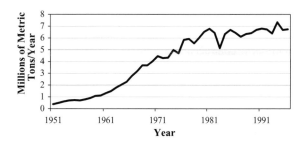

Fig. 2. Annual nitrogen input from fertilizer, Mississippi–Atchafalaya River Basin (Goolsby et al., 1999).

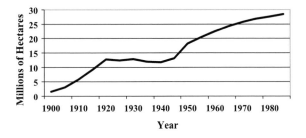

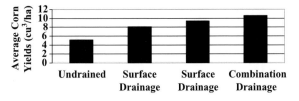

Fig. 3. Land area drained in the Mississippi–Atchafalaya River Basin (Mitsch et al., 1999).

Fig. 4. Effect of drainage intensity on corn yield (Zucker and Brown, 1998).

watershed. By the 1980s, more than 28 million ha (70 million acre) in the Mississippi Basin had been drained by extensive systems of drain tiles and outlet ditches (Fig. 3). These systems were constructed to move surface and groundwater out of agricultural fields efficiently. Field drainage directly impacts corn yield (Fig. 4). A hectare (acre) without drainage might yield $2.1\,m^3$ (60 bushels) of corn, whereas the same hectare

with a properly designed surface drainage system could yield $3.2\,m^3$ (90 bushels) of corn (Zucker and Brown, 1998).

Many, but not all, of the drained areas had once been wetlands. Not surprisingly, as drainage has increased, total wetland area in the basin has decreased. Dahl (1990) estimated that six Upper Mississippi River Basin states—Iowa, Missouri, Illinios, Indiana, Ohio, and Kentucky—lost 80–90% of their wetlands from 1780 to 1980 (Fig. 5). This is approximately the same region with the highest fertilizer usage (Fig. 6). Conse-

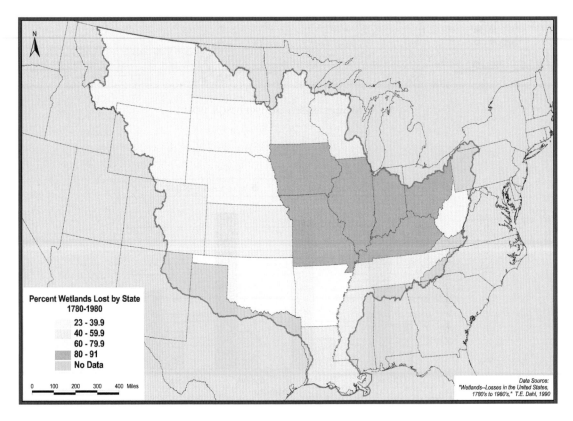

Fig. 5. Percentage of wetlands lost, 1780–1980 (Dahl, 1990).

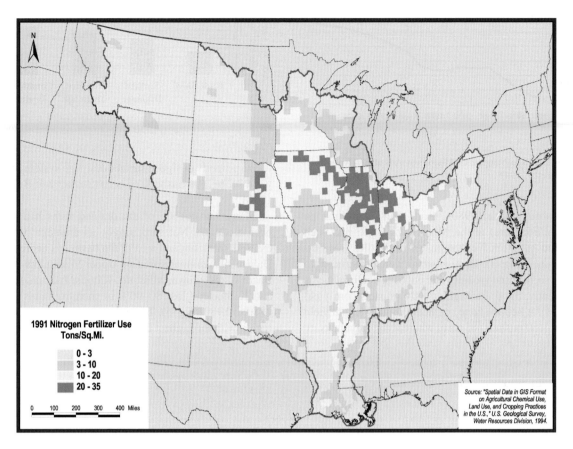

Fig. 6. Nitrogen fertilizer use, 1991 (USGS, 1994).

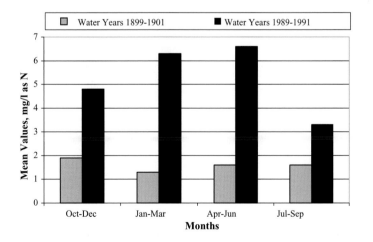

Fig. 7. Nitrite–nitrate concentrations in the Illinois River near Peoria (Palmer, 1902; USGS, 1989–1991).

quently, these regions typically have streams and rivers with highly elevated nitrite–nitrate concentrations due to both non-point and point discharges. For example, the nitrite–nitrate concentrations in the Illinois River near Peoria have changed in magnitude and distribution. Mean concentrations have increased from 1.9 to 6.5 mg/l over the past 100 years. At the same time, the highest seasonal concentrations shifted from fall to spring (Fig. 7).

Past efforts have failed to persuade farmers to reduce fertilizer usage and limit runoff from their fields. The reason for these failures is simple: the farmer was not been adequately compensated. For example, crop yield reduction would not be offset by an increase in price. The farmer clearly understands that fertilizers and drainage affect income, but designers of past regulatory programs have not realized this economic reality.

3. Flooding

Mean annual national flood damages have been steadily increasing; now approximating US$ 3.4 billion (Fig. 8). This rise has occurred despite a similar rise in control costs (Fig. 9). The need for flood storage was never more apparent than during the 1993 floods on the Mississippi River and its tributaries when flood losses were US$ 16 billion (Richards, 1994). Floodwalls at St. Louis would have been overtopped if numerous levees upstream had not failed, allowing the accumulated floodwaters to inundate former floodplain areas. At the time, owners of failed levees likely gave little

thought to the economic value of their flooded property as a place to store floodwaters. Had upstream levee districts been organized and downstream owners well informed, a market for floodwater storage certainly would have flourished in the late spring and summer of 1993.

Avoiding these losses would have necessitated an additional 49 billion m^3 (40 million acre-ft) of flood storage. Based on resulting damages from the 1993 flood, a cubic meter of storage would have been worth US$ 0.33. Assuming that the floodwaters had been stored in shallow (less than 1 m deep) wetlands located behind the levees, 5.3 million ha would have been required. The rental rate for the 1993 event would have yielded US$ 486/ha, a handsome sum, since most agricultural land in the Midwest rents for US$ 40–60/ha. If floodwaters were stored in deeper pools behind levees, the unit-area rental rate would be even higher. The probability of a flood equaling the magnitude of the 1993 occurrence is 0.01 (1%) in any given year, so the projected value of damages would be approximately US$ 160 million (US$ 16 billion × 0.01). This would result in an annual value of only US$ 5/ha for a wetland flood-storage credit. If floodwater could be stored to a depth of 4.5 m, however, the value would be US$ 24/ha/year. This income would be in addition to the normal income from farming or other activities that would only rarely be interrupted.

4. Solution

The scale of wetland restoration needed to solve current nutrient, sediment, and flood control problems is

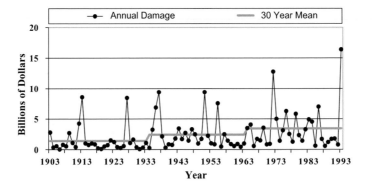

Fig. 8. National Annual Flood damages, 1993 dollars (Richards, 1994).

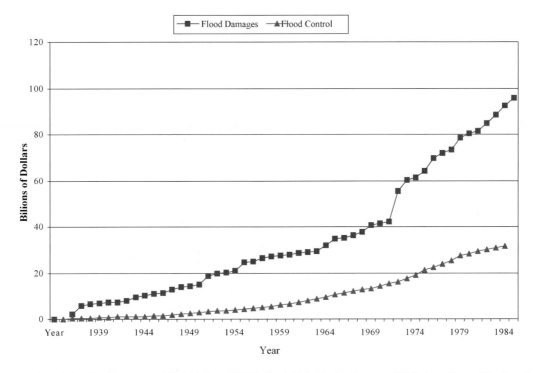

Fig. 9. Cumulative Flood Damage and Control Costs (1985 Dollars) (U.S. Weather Bureau and U.S. Army Corps of Engineers).

enormous. In the Mississippi River basin, for example, an estimated 2–5.3 million ha of restored wetlands would be required to stem the flow of nitrogen to the Gulf of Mexico, which, in turn, would diminish the hypoxic zone (Mitsch et al., 2001). The job is too enormous to be completed solely by government grants; market-based solutions are needed.

Our proposed strategy, called "nutrient farming," could overcome the lack of economic incentive. This strategy involves converting lowlands from corn production to that derived from wetlands. The wetland accepts the inflow of nitrogen-laden water from upstream and discharges low-nitrogen water downstream. Through the inherent biogeochemical and physical processes in the wetland, a high percentage of nitrogen and phosphorus would be removed, depending on the location and size of the wetland. The farmer would record the amount removed, which would then be certified by the state. This is similar to the current self-monitoring practice used by operators of water reclamation plants who must provide daily monitoring reports to federal and state regulatory agencies. Once the

farmer received state certification, he/she would be free to sell certified nitrogen credits on the open market or through long-term contracts. Buyers of nitrogen credits might be other farmers, municipal and industrial dischargers, or other industries that release nitrogen and phosphorus to the atmosphere or aquatic ecosystems through point or non-point emissions (agriculture, automobile, manufacturing, etc.). Nutrient farming could create a market where the services of nutrient farming could be bought and sold, thereby creating a whole new economic structure for farming that focuses on optimal resource allocation and economic efficiency. Nutrient farming provides an alternative income for farmers.

Based on a preliminary investigation, the cost of producing a metric ton of nitrogen credit at a wetland nutrient farm would be approximately US$ 2500 (Hey et al., 2005). This covers the costs of the land and wetland restoration, as well as the labor and materials needed to operate the nitrogen farm. The true cost of corn production should include a per metric ton surcharge on applied fertilizer because the drainage removes approx-

Table 1
Total annual cost comparison between wetlands and water reclamation plant (WRP) treatment (Hey et al., 2005)

Criteria Limit	Wetland Area (ha)	Total annual cost (US$)				Total annual cost with sale of excess credits (US$)			
		Wetland total	WRP total	Savings	%[a]	Wetland total	WRP total	Savings	%[a]
3.0 mg/l TN 1.0 mg/l TP	76,500	63,900,000	174,000,000	110,000,000	63	40,900,000	174,000,000	133,000,000	76
2.18 mg/l TN 0.5 mg/l TP	130,000	103, 000, 000	211,000,000	108,000,000	51	46,000,000	211,000,000	164,000,000	78

[a] Percent savings compared to WRP total annual costs.

imately one-third of every metric ton of fertilizer applied to a crop. Therefore, a farmer should be required to purchase nitrogen credits for that third of a metric ton sent downstream. This could add US$ 830 (one-third of the US$ 2500 cost to remove nitrogen) to the farmer's per metric ton cost of fertilizer (currently US$ 181).

Considering just the potential market available from farmers buying credits for their fertilizer usage, the market size is considerable. In 1998, for example, United States farmers purchased an estimated 11.3 million Mt of nitrogen fertilizer (FAO, 2002) at US$ 181/Mt (TFI, 1960–2000) for a total cost of US$ 2 billion. Assuming that one-third of the fertilizer entered the drainage system, then almost 4 million metric tons of nitrogen credit would have been required to offset the load on a 1:1 basis. The value of these credits would total US$ 10 billion annually (4 million metric tons of credit × US$ 2500/Mt removal).

Another, more immediate nitrogen credit market exists for industrial or municipal point source dischargers. To meet USEPA criteria for nitrogen and phosphorous, U.S. wastewater treatment plants must be upgraded. The cost of conventional treatment technology to remove nutrients to near-criteria levels was estimated for the Illinois Association of Wastewater Agencies (IAWA) (Zenz, 2003). The capital cost was estimated at US$ 5.3 billion for the 814 publicly owned treatment works (POTWs) in Illinois. Annual operation and maintenance costs were estimated at nearly US$ 500 million. Based on these numbers, capital costs nationwide would reach about US$ 26 billion with annual operations and maintenance costs at US$ 4.8 billion. If wastewater agencies would be willing to spend up to the estimated operations and maintenance costs to avoid the capital costs, the market in nitrogen and phosphorous credits could reach US$ 4.8 billion/year.

Based on a cost comparison analysis performed between conventional biological nutrient removal at wastewater treatment facilities and treatment wetlands, nutrient farms can cost effectively remove nitrogen to proposed nutrient criteria (Hey et al., 2005). For example, the annual cost for a sanitary district located in the upper Illinois River watershed to construct and implement biological nutrient removal control to meet nutrient criteria of 2.18 mg/l TN and 0.5 mg/l TP would be US$ 211 million (Table 1). The annual cost of restoring and operating 130,000 ha (322,000 acres) of nutrient farm wetlands, which is the area required to remove the sanitary district's monthly demand, is US$ 103 million or 51% less than advanced wastewater treatment costs. Since the performance of treatment wetlands is seasonally dependent, the nutrient farms have to be designed to meet the monthly demand of industrial and/or municipal dischargers. Therefore, during certain times of the year, such as the summer or fall, there is an excess capacity to remove nutrients in the wetlands and the nutrient farms could generate surplus credits. Thus, the 130,000 ha (322,000 acres) of treatment wetlands would remove a surplus of 26,100 Mt (28,800 t) of nitrogen and 2000 Mt (2200 t) of phosphorus with a total value of approximately US$ 56.3 million. If secondary markets for these excess nutrient credits could be developed, then the savings could reach as high as 60–70% of the cost of conventional biological nutrient removal.

The market area for nitrogen farming would be restricted to the more humid, eastern portion of the United States and along the Pacific coast where row crops are intensely grown. The western portions of the Missouri River basin produce little nitrogen-ladened runoff due to low precipitation and the particular crops grown in the region. On the other hand, rivers along coastal California and rivers draining southern Minnesota, Iowa,

Missouri, and Arkansas and all states eastward would present fertile areas for nitrogen farming. The magnitude of each market could be estimated from fertilizer sales, emissions of power plants and automobiles, and the discharge of municipal and industrial point sources.

The cornerstone for the nitrogen market has been laid. In January 2001, the United States Environmental Protection Agency promulgated nutrient criteria, including nitrogen, for the streams and rivers of the nation. In the "Cornbelt Eco-region," where most of the nitrogen fertilizer is used, the total nitrogen criterion was set at 2.18 mg/l. Of this concentration, about 1.6 mg/l is nitrate–nitrogen. This is substantially lower than the nitrogen concentrations of now conveyed by Midwestern rivers: Peak concentrations of nitrate–nitrogen in the Sangamon River in central Illinois have reached 12–16 mg/l over the past decade. For the Illinois River, approximately 90,700 Mt/year will need to be removed. At the estimated cost for credit production of US$ 2500/Mt, the annual market for nitrogen credits would be approximately US$ 227 million.

The nitrogen market would need to be locally based. Dischargers of nitrogen would likely have to buy credits from sellers upstream of their discharge point. In this way, elevated concentrations of nitrogen in the river system would be minimized. In cases where the nutrient farm is located at a distance from the point of nutrient discharge, regulatory agencies will need to identify stream reaches where nutrient transport will be allowed as a designated use. These reaches will likely be in highly modified channels (e.g., the Illinois Waterway) and removed from primary contact and drinking water uses. This reclassification, however, would require a change in the way water quality standards are currently administered. In many cases, nutrient farms will be able to be sited near the point of discharge; therefore, transport issues will not be of concern.

The nitrogen credit market would remain viable as long as crops need to be grown on well-drained and fertilized land. A shift to crop species more tolerant of saturated soils or a change in the chemical nature and application of fertilizer could reduce the need for such credits. Currently, crops requiring less drainage and nitrogen are being actively considered. New forms of nitrogen fertilizer that attach to soil and dissolve more slowly are being developed. Nonetheless, current market opportunities are enormous.

To expand the nutrient market, both phosphorus and carbon should be considered. The very wetlands that act as nitrogen farms also would act as phosphorus and carbon sinks. Credits could be sold to emitters of carbon to water and air. Consequently, nitrogen farms could be used to reduce greenhouse gases. Finally, a sediment market could be structured in the same way as the other credits. Sequestering sediments in wetlands would protect downstream aquatic habitats and reduce the dredging costs incurred by the U.S. Army Corps of Engineers that are related to maintaining river navigation.

References

Dahl, T.E. 1990. Wetland losses in the United States: 1780–1980s. US Department of the Interior, Washington, DC.

FAO 2002. Fertilizer use by Crop, 5th ed. Joint publication by Food and Agriculture Organization of the United Nations (FAO), International Fertilizer Industry Association, the International Fertilizer Development Centre, the Phosphate and Potash Institute, and the International Potash Institute. FAO, Rome. Viewed (9/20/04) at http://www.fertilizer.org/ifa/statistics/crops/fubc5ed.pdf.

Goolsby, D.A., Battaglin, W.A., Lawrence G.B., Artz, R.S., Aulenbach B.T., Hooper, R.P., 1999. Flux and sources of nutrients in the Mississipp–Atchafalaya Basin: Topic 3 Report for the Integrated Assessment on Hypoxia in the Gulf of, Mexico, Silver Spring (MD), NOAA, Coastal Ocean, Office, Decision Analysis Series no. 17.

Hey, D.L., Kostel, J.A., Hurter, A.P., Kadlec, R.H., 2005. Comparative economics of nutrient management strategies: traditional treatment and nutrient farming. Water Environment Research Foundation Report #03-WSM-6CO, Alexandria, VA.

Hoeft, R.G., Nafziger, E.D., Mulvaney, R.L., Gonzini, L.C., Warren, J.J., 1999. Effect of time and rate of N application on N use efficiency and surface water contamination with nitrates. In: Hoeft, R.G., (ed), 1999 Illinois Fertilizer Conference Proceedings, pp. 39–54.

Johnson, B.N., Baumgart, P., Kent, P., Kramer, J.M., Tutwiler, A.B.F., 2001. Phosphorus Credit Trading in the Fox–Wolf Basin: Exploring Legal, Economic and Technical Issues. Water Environment Federation, Alexandria, Virginia.

Mitsch, W.J., Day Jr., J.W., Gilliam, J.W., Groffman, P.M., Hey, D.L., Randall, G.W., Wang, N., 1999. Reducing nutrient loads, especially nitrate–nitrogen, to surface water, ground water, and the Gulf of Mexico. Topic 5 Report. Available from http://www.nos.noaa.gov/products/pubs_hypox.html.

Mitsch, W.J., Day Jr., J.W., Gilliam, J.W., Groffman, P.M., Hey, D.L., Randall, G.W., Wang, N., 2001. Reducing nitrogen loading to the Gulf of Mexico from the Mississippi River Basin: Strategies to counter a persistent ecological problem. Bioscience 51 (5), 373–388.

Mitsch, W.J., Gosselink, J.G., 2000. Wetlands, 3rd ed. Wiley, New York.

Moore, R.E., Overton, M.S., Norwood, R.J., DeRose, D., Corbin, P.G., 2000. Nitrogen Credit Trading in the Long Island Sound Watershed. Water Environment Federation, Alexandria, Virginia.

Palmer, A.W., 1902. Chemical survey of the waters of the Illinois. University of Illinois, Champaign-Urbana.

Rabalais, N.N., Turner, R.E., Scavia, D., 2002. Beyond science into policy: Gulf of Mexico hypoxia and the Mississippi River. Bio-Science 52, 129–142.

Richards, F. 1994. U.S. Weather Bureau. Personal Communication.

TFI (The Fertilizer Institute), 1960–2000. Average U.S. farm prices of selected fertilizers, Agricultural Prices, National Agricultural Statistics Service, USDA. Viewed (9/20/04) at www.tfi.org/statistics.

U.S. Geological Survey, 1989. Water Resources Data, Illinois: Water Year 1989, vol. 2. Illinois River Basin. U.S. Geological Survey Water Data Report, IL-89-2, Urbana, IL.

U.S. Geological Survey, 1990. Water Resources Data, Illinois: Water Year 1990, vol. 2. Illinois River Basin. U.S. Geological Survey Water Data Report, IL-90-2, Urbana, IL.

U.S. Geological Survey, 1991. Water Resources Data, Illinois: Water Year 1991, vol. 2. Illinois River Basin. U.S. Geological Survey Water Data Report, IL-91-2, Urbana, IL.

U.S. Geological Survey, Water Resources Division. 1994. Spatial data in GIS format on agricultural chemical use, land use, and cropping practices in the U.S.

Weyer, P.J., Cerhan, J.R., Kross, B.C., Hallberg, G.R., Kantaneni, J., Breuer, G., Jones, M.P., Zheng, W., Lynch, C.F., 2001. Municipal drinking water nitrate level and cancer risk in older women: The Iowa women's health study. Epidemiology 12 (3), 327–338.

Zenz, D.R. 2003. Technical feasibility and cost to nutrient standards in the state of Illinois. Report commissioned by the IL Assoc of Wastewater Agencies.

Zucker, L.A., Brown, L.C. (Eds.), 1998. Agricultural Drainage: Water Quality Impacts and Subsurface Drainage Studies in the Midwest. Ohio State University Extension, Bulletin 871. The Ohio State University, 40 pp.

Available online at www.sciencedirect.com

Ecological Engineering 24 (2005) 289–307

ECOLOGICAL ENGINEERING

www.elsevier.com/locate/ecoleng

ELSEVIER

Hambleton Island restoration: Environmental Concern's first wetland creation project

Edgar W. Garbisch *

Environmental Concern Inc., 201 Boundary Lane, P.O. Box P, St. Michaels, MD 21663, USA

Received 13 August 2003; received in revised form 9 September 2004; accepted 1 November 2004

Abstract

Early in 1971, the author began a 1-year sabbatical near Hambleton Island (HI) after resigning from his teaching/researching chemistry professorship at the University of Minnesota in Minneapolis. A new career for the author and this project commenced after reading "The Life and Death of a Salt Marsh" by Teal and Teal [Teal, J., Teal, M., 1969. Life and Death of the Salt Marsh. Little, Brown and Co., Boston, MA]. After obtaining the necessary Maryland and Army COE permits to proceed with this project, one of the author's chemistry post doctorates, Dr. Paul Woller, and one of his graduate students who just received his Ph.D. in chemistry, Dr. Robert McCallum, joined the author to restore a section of Hambleton Island that had eroded into two islands, back to a single section through the creation of over 0.8 ha of tidal brackish marsh. This is a story of the creation of a tidal marsh by three Ph.D. chemists who had never grown a plant and, at the beginning, knew nothing about wetlands. In 1972, the author formed Environmental Concern Inc., a public not-for-profit corporation that has constructed (created, restored, or enhanced) over 700 non-tidal and tidal wetlands throughout mostly the eastern USA through year 2003.
© 2005 Elsevier B.V. All rights reserved.

Keywords: Upland shore erosion; Tidal brackish marsh; Canada geese; Peat bank; Sediment entrapment; Litter/debris deposits

1. Introduction

Fig. 1a shows the Chesapeake Bay and the location of the work site, which includes Hambleton Island (HI). In 1971, Hambleton Island (HI) (at the confluence of Broad Creek, San Domingo Creek, and Edge Creek on the Eastern Shore of the Chesapeake Bay near the town of St. Michaels, MD) consisted of three rapidly eroding

upland forested islands (see Fig. 1b). Earlier records show that HI was a single (22 ha) agricultural island in 1849 that supported two residences. When the agriculture on the island ceased, HI became forested and any shoreline marsh vegetation was shaded out, leading to increased upland erosion rates. By 1939, HI had eroded into two islands totaling 12 ha. Around 1969, the largest of the two islands had eroded into two islands. By 1971, these two islands were separated by a breach approximately 30 m wide and HI had decreased in area to 6 ha (Fig. 2).

---

* Tel.: +1 410 745 9620; fax: +1 410 745 3517.
  E-mail address: E.Garbisch@wetland.org.

0925-8574/$ – see front matter © 2005 Elsevier B.V. All rights reserved.
doi:10.1016/j.ecoleng.2004.11.013

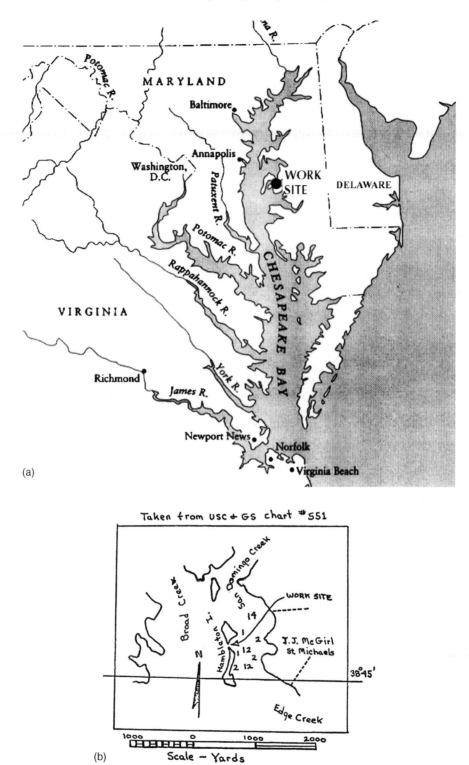

Fig. 1. (a) Chesapeake Bay and the work site (HI) and (b) sketch of the work site as seen on the permit application to the Baltimore District COE in 1971.

Fig. 2. A 1971 photograph of HI showing the three islands and the breach.

2. Results and discussion

The prevailing winds at the project site during the winter months are from the west/northwest and during other times of the year they are from the south/southwest. As seen in Fig. 1b, the west side of HI is most vulnerable towards erosion. The major project objective was to create over 0.8 ha of tidal wetlands at the work site in Figs. 1b and 2, thereby restoring the two sections of HI at the breach to a single section. It was expected that as this single section of HI continued to erode from west/northwest and south/southwest winds, that the suspended eroded soils would be carried through the breach and trapped/collected by the created wetland. Thus, as the upland of this section of HI reduced in area due to erosion, the wetland was expected to increase in area.

The mean tidal range at HI is 0.43 m and the high spring tidal range is 0.76 m. The depth of water throughout the work site area at mean low water ranged from 0 to 0.3 m. The water salinity at HI ranges between 8 and 20 ppt. The lower salinity is seen in the spring and following periods of heavy rain add the higher salinity is found during periods of drought. Because the water salinities are lower than those of coastal salt marshes, vegetation communities in brackish tidal marshes generally have higher species diversity than found in salt marshes (Tiner and Burke, 1995; Sipple, 1990).

The author obtained a permit from the Baltimore District Army Corps of Engineers in 1971 to restore the two recently formed sections of HI back to a single section by constructing 0.8 ha of tidal brackish marsh at the work site (Figs. 1b and 2). Biological benchmarks (Garbisch et al., 1975; Garbisch, 1986, 1989; Denbow et al., 1996) were used to estimate mean high water (NHW) at the project site. The MHW elevation then was tied into a physical benchmark on HI. The tidal range at St. Michaels, MD, San Domingo Creek, is given in the annual NOAA tide tables of the east coast of North and South America.

The design was to construct 0.8 ha of tidal marsh at the mid-tide elevation at the work site and to grade to the top of the upland bank at HI on an approximate 10:1 slope. In order to achieve this design, 500 barge loads of clean (sterile) quarry material (at 6 M^3 per load for a total of 3000 M^3) had to be moved to the work site at HI at and about high tide during November and December of 1971 and January through June of 1972. The quarry material consisted of 4.4% gravel, 79.3% sand (primarily medium sized), and 16.3% mud (silt and clay)(Garbisch et al., 1975). The draft of the loaded barge was too great to get the barge into the

Fig. 3. (a) Loading barge with 6 M^3 of medium sized sand and (b) barging sand to HI, approximately 3 km from loading location, at and around high tide during the periods November–December 1971 and January–June 1972.

work site at tides lower than those from 2 h before and after the times of high tide (Fig. 3). Sand was chosen over finer grained soils because sand could be graded most precisely and water turbidity would be minimized during the offloading and grading operations, thereby minimizing any water quality problems arising from high water turbidity.

Once at the work site, the sand was washed off of the barge throughout pre-staked areas using a high-

pressure pump (Fig. 4a and 4b). During times of low tide, the sand was fine-graded using a high-pressure pump (see Fig. 4c) to a sand flat at mid-tide sloping up to the top of the upland banks on an approximate 10:1 slope (Fig. 5).

Originally, it was intended to protect the entire work site with plastic breakwaters in order to ensure that the sand base for the constructed wetland was stable until the marsh vegetation became well established.

Fig. 4. Washing sand off of the barge throughout pre-staked areas using a high-pressure pump: (a) January 1972 and (b) June 1972. (c) Fine-grading at low tide. Note slope to upland in the background that has been planted.

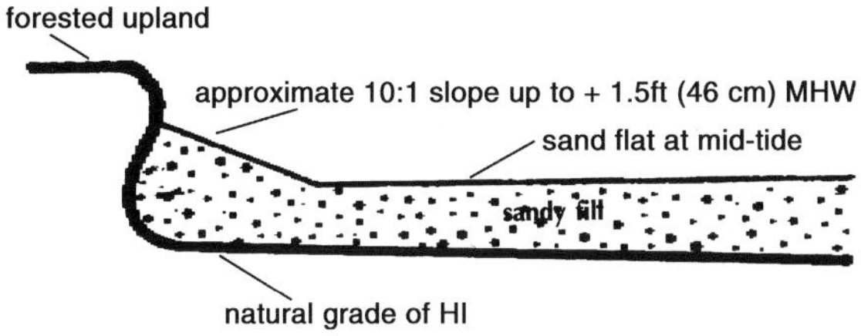

forested upland

approximate 10:1 slope up to + 1.5ft (46 cm) MHW

sand flat at mid-tide

sandy fill

natural grade of HI

Fig. 5. Poor quality photo of sand flat at mid-tide elevation, graded on an approximate 10:1 slope to the top of the upland bank.

However, as work progressed and the wave dynamics throughout the work site became better understood, it was found necessary to protect only the highly exposed area at and just west of the breach (Figs. 1b and 2), Fig. 6 shows the plastic breakwaters that enclose the area just west of the breach at HI. The footings of these breakwaters rapidly became buried with sediment and the breakwaters could not be removed after they were no longer needed by 1974. By 1977, however, most of the breakwaters had been destroyed by ice, ultraviolet light, and floating debris.

Over 60,000 peat potted herbaceous plants, ranging from six to 16 weeks in age, were grown and planted to the created sand areas at HI in 83 monitoring plots during the period of 1 April to 16 September 1972. The monitoring program, which was the pursuance of the scientific work, began at the time of planting. A total nine herbaceous species were grown and planted (Table 1). All of the species are wetland plants with the exception of American beachgrass, which is an upland dune plant and planted at the upland elevations of the created shore (see Fig. 5). Five of the species were planted at elevations from mean high water (MHW) to the high spring-tide. This area constitutes less that 5% of the total planted area. Two *Typha* species were planted at the higher end of the normal tidal range. Broad-leaved cattail (*Typha latfolia*) was not expected to survive due to its low salt tolerance. Over 95% of the total planted area contained smooth cordgrass (*Spartina alterniflora*) at evaluations from mid-tide to MHW. All of the plants were grown from seed collected from areas given in Table 1. None of the plants were conditioned to be planted into the prevailing water salinities of 8–20 ppt. The growing of the plants in Table 1 was the beginning of the first native wetland plant wholesale nursery in the USA.

Fig. 7a shows the work site in 1971 prior to the start of work. Fig. 7b shows the same view of the work site

Fig. 6. Plastic breakwaters protecting the breach area until plants are well-established.

Table 1
Species planted to created wetlands at Hambleton Island, Maryland in 1972

Botanical name (common name)	Seed source[a]	Age[b]	Planting period[c]	Elevation planted[d]	Salinity tolerance
Ammophila breviligulata (American beachgrass)	A	8–12	May 13 to June 8	Above storm tide elevation	Less that 1 ppt
Distichlim spicata (salt grass)	A	10–15	April 3 to July 5	MHW to high spring-tide	Up to 50 ppt
Panicum virgatum (switchgrass)	C	7–12	April 19 to May 13	MHW to high spring-tide	Up to 10 ppt
Phragmites australis (common reed)	C	11–16	April 18 to Sept 16	MHW to high spring-tide	Up to 20 ppt
Spartina alterniflora (smooth cordgrass)	A, B	7–15	April 18 to August 3	Mid-tide to MHW	Up to 35 ppt
Spartina cynosuroides (big cordgrass)	A	8–14	April 11 to August 30	MHW to high spring-tide	Up to 10 ppt
Spartina patens (saltmeadow hay)	A	6–13	April 12 to July 5	MHW to high spring-tide	Up to 35 ppt
Typha angustifolia (narrow-leaved cattail)	C	8–13	April 1 to July 2	Upper 20% of the mean tide range	Up to 15 ppt
Typha latifolia (broad-leaved cattail)	C	8–13	April 1 to August 30	Upper 20% of the mean tide range	Less than 1 ppt

[a] Key to geographical locations from which seed was obtained: A, Assateague Island, VA; B, coast of North Carolina; C, St. Michaels, MD.

[b] Length of time in weeks that seedlings were under cultivation in greenhouse prior to planting.

[c] In 1972.

[d] Normal tidal range at HI is 0.43 m. High spring tide range is 0.76 m. Mean low water elevation of 0.0 m.

in May of 1972 after the site had been filled with sand, graded, and partially planted. Fig. 8 is an aerial view of the work site in July of 1972. By this date, planting of Site A had been 98% completed. The area landward of the hashed white line (Site B) was filled with upland stockpiled and sterile dredged materials and planted in 1973. Fig. 7c is the same view of the work site as in Fig. 7a and b, but in September of 1972 showing the smooth cordgrass in flower.

All of the seedlings utilized for this project resulted from incubator germinated seed. After germination, a known number of seedlings were hand planted into sand-filled 5.7 cm peat pots. After cultivation in a greenhouse for 6–8 weeks, the seedlings were either transplanted to the project site or transplanted into sand-filled 10 cm peat pots and cultivated for an additional 2–8 weeks before being transplanted to the project site. Seedlings were normally transplanted to the project site 0.9 m on center; however, 0.3 m, 0.46 m and 0.6 m on center spacings were on occasion used. Fertilization was or was not accomplished at the time of transplantation.

With the exception of *T. latifolia*, which did not survive due to its intolerance to the water salinity and of *Typha angustifolia,* which had a low survival per-

centage of 13% due to its intolerance to wave stress and its being continually grazed during the growing season, all of the seven other species had excellent survival. Over 95% of the site consisted of *Spartina alterniflora* transplants, which had an overall survival of 95% during the first growing season. The survival of seedling transplants of *Spartina patens* and *Daistichlis spicata* was 100%, and that of *Panicum virgatum* and *Ammophila breviligulata* was 97%, and that of *Spartina cynosuroides* and *Phragmites australis* was 75%.

Macrobenthic invertebrates were monitored, from 5 cm below MLW to MHW at Sites A and B (see Fig. 8) and at the control site (Site C in Fig. 2) starting in June 1973, 3 months after the filling and grading of Site B with sterile dredged materials consisting of 58% sand (primarily fine to very fine) and 41.9% mud (silt and clay). After 1 year of monitoring, all three sites contained statistically different total populations of macrobenthic invertebrates with Sites B and A having the highest and lowest populations, respectively (Garbisch et al., 1975). The total invertebrate populations increased significantly with decreasing elevation from MHW to 15 cm below MLW (Garbisch et al., 1975).

Fig. 7. View of the work site: (a) in 1971 prior to the start of work; (b) in May 1972 following the development and partial planting of the wetland; and (c) in September 1972 with the smooth cordgrass in flower.

In early April 1973, a flock of Canada geese (*Branta canadensis*) flew in to the wetland on a moonlit night during a high tide and fed on the marsh for several hours. The marsh was severely eaten-out as a result of this feeding (Fig. 9). As a result of this eat-out, the monitoring program and all of the planned research abruptly terminated. Fig. 10 shows the layout of all of the 83 monitoring plots of different species of different plant ages and plant spacings, different wave climate exposure, and different fertilization regimes. The

Fig. 7. (*Continued*).

monitoring plot stakes are shown in Fig. 9 with most of the associated vegetation eaten-out.

Geese feeding took place after all of the smooth cordgrass had flowered (see Fig. 7c) and its seed shat-tered, and their feeding effectively sowed its seed. Consequently, only smooth cordgrass seedlings were seen throughout the intertidal sand flat in May of 1973.

Fig. 8. Aerial view of the work site in July 1972. The area landward of the white hashed line, Site B, was filled with dredged materials and planted in 1973. The darker rows of plants throughout Site A have been growing on site for a longer period than the lighter rows.

Fig. 9. (a) Inspecting the site in early April 1973. (b) A view of the damage caused by the geese. Note the monitoring plot stake in the foreground. The geese feed on marsh plants during the dormant season by washing the sediment away from their root zones during times that water covers the marsh. They then feed on the exposed belowground perennial parts, leaving behind the excavation holes seen in (b). (c) The many monitoring plot stakes in the absence of vegetation.

Fig. 9. (*Continued*).

The Canada geese incident quickly led to an effective and inexpensive method of protecting herbaceous wetland plants from Canada geese. It was noted that Canada geese do not fly and land into fairly mature stand of plants. They fly and land into open water and then swim into the plants and start feeding. During the growing season, geese feed on the aboveground plant parts. During the dormant season, as in this instance, the geese wash (with their feet) the sediment away from the belowground plant parts after the tide covers the substrate and they then feed on these exposed plant parts. Excluding the geese from the plants is a method to avoid eat-outs. Fig. 11a shows the erection of a chicken wire goose exclosure fence at the work site following the eat-out. It was quickly determined that the less expensive wooden stake/cloth line exclosure fence shown in Fig. 11b was equally effective. Lines are placed every 15 cm, starting 15 cm above the sediment surface and

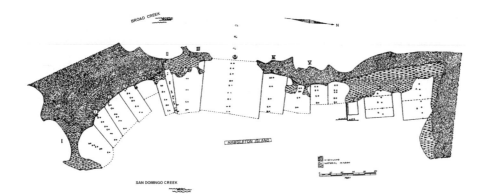

Fig. 10. The layout of the 83 experimental monitoring plots at the work site. Canada geese terminated the experimental component of the project the evening that they ate-out the marsh.

Fig. 11. Canada geese exclosure methods: (a) erecting a chicken wire Canada goose exclosure fence at the work site at HI; (b) a wooden stake/cloth line Canada goose exclosure fence.

Fig. 12. View of the smooth cordgrass at the work site in October 1973.

ending 15 cm above the high water level at the site. If the area of exclusion is large enough for Canada geese to fly into, when the vegetation is immature, it is recommended to install stakes with flags attached throughout the planted area to obstruct geese from flying in.

In May 1973, the intertidal sand flat was dominated by smooth cordgrass seedlings. By October 1973, the smooth cordgrass had matured and flowered (Fig. 12). The view in this figure is the same as in Fig. 7, which were seen in 1972. Note the loss of trees from upland erosion in less than a year.

The same aerial views of the work site at HI in 1971 and in 1976 are shown in Fig. 13. In the 1976 photograph, litter/debris deposits can be seen landward of the breach. These deposits are more clearly seen in the 1977 photograph shown in Fig. 14. It was expected that, as the main section of HI continued to erode, the eroded sediments would be washed through the breach by the prevailing winds and be trapped by the marsh, increasing its area as the upland area decreased by erosion. As it turned out, the marsh was so effective in trapping the eroded sediments that the elevation at the breach rapidly increased to 0.6 M above MHW, thereby

preventing all waters carrying the eroded sediments of HI from passing through the breach except those during storm high tides. Consequently, the deposits seen in Fig. 14 would eventually wash over the rest of the marsh during periods of storm high tides.

Fig. 15 shows the same view of the work site September 1977 and August 1992 and Fig. 16 shows the same view of the work site in October 1973 and August 1992. Note the extensive loss of trees as the time passed. Also note that the common reed (*Phragmites australis*) that was planted in small areas from MHW to high spring tide (see Table 1) dominated the elevations above MHW throughout the breach and elsewhere by 1992 (see Figs. 15 and 16). This dominance by the aggressive common reed led to the disappearance over time of the other four species (i.e., salt grass (*Disstichlis spicata*), switchgrass (*Panicum virgatum*), big cordgrass (*Spartina cynosuroides*), and saltmarsh hay (*Spartina patens*)) that were planted at and above MHW at HI (see Table 1) in 1972 and to a constructed tidal brackish marsh having lower species diversity than would have been expected had common reed not have been planted (Tiner and Burke, 1995; Sipple, 1990).

1971   1976

Fig. 13. Aerial views of the work site before the start of work in 1971 and after the work was completed in 1976.

Fig. 14. Debris and litter deposits near the breach at HI in 1977.

Fig. 15. View, looking north, across the tidal marsh: (a) in September 1977; (b) in August 1992.

By 1992, a peat bank 0.3 m above the original grade had developed throughout the marsh from the belowground biomass production of the smooth cordgrass (see Fig. 17). This increased elevation of the smooth cordgrass marsh normally would have seen it succeed to another higher elevation plant species. The tidal range at HI is 0.43 m and smooth cordgrass was planted throughout the elevations of mid-tide 0.21 m to MHW 0.43 m. However, by 1992, the depth of water on the marsh surface at the time of MHW was essentially the same as it was in 1972. Consequently, the combination of sea level rise and land subsidence at the work site during the period of 1972 through 1992 led to the same effective rise in water level as the rise in marsh surface through peat development. Horton (2003) discusses land sub-

Fig. 16. View, looking south, across the tidal marsh (a) in October 1973 and (b) in August 1992.

sidence and sea level rise throughout the Chesapeake Bay.

Fig. 18 is an aerial photograph of HI in the winter of 2002. Note the dramatic loss of upland due to erosion when compared to the 1971 photograph of HI shown in Fig. 2. Also note in Fig. 18 the light-colored footprint

of the original HI, years back before it started eroding to its present condition. Fig. 19a is a winter 2002 aerial photograph of the created wetland, and is the same view as that taken in the fall of 1976 and shown in Fig. 19b. Note the peat bank edge of the wetland and the muskrat runs into the wetland in Fig. 19a.

Fig. 17. Smooth cordgrass in the breach area of HI in 1992 that shows the 0.3 m high peat bank that developed over the 20-year time period. Note the common reed in back of the smooth cordgrass (top right) that moved into the high elevation sediment deposit area near the breach from nearby high elevation areas where it was originally planted.

Fig. 18. Aerial view of HI during the winter of 2002.

Fig. 19. Aerial view of the tidal marsh: (a) during the winter of 2002 and (b) 26 years earlier in the fall of 1976.

3. Conclusions

Although partial accounts of the Hambleton Island restoration have appeared elsewhere (Cantor, 1997; Garbisch, 2002), this is the first complete report of observations through year 2002. It is a special project for the author. Even though the great majority of the scientific research work was instantly aborted by the Canada geese eat-out in the early spring of 1973, it opened the way for a second career in all aspects of wetland construction and for the creation of the first wholesale wetland nursery in the United States.

References

Cantor, S.L., 1997. Contemporary Trends in Landscape Architecture. Environmental Concern Inc., pp. 11–20. Van Nostrand Reinhold, New York, NY, 348 pp.

Denbow, T.J., Klements, D., Rothman, D.W., Garbisch, E.W., Bartodous, C.C., Kraus, M.L., MacClean, D.R., Thunhorst, G.A., 1996. Guidelines for the Development of Wetland Replacement Areas. National Cooperative Highway Research Program. Report 379. National Academy Press, Washington, DC, 79 pp.

Garbisch Jr., E.W., Woller, P.B., McCallum, R.J., 1975. Salt Marsh Establishment and Development. TM 52. U.S. Army Corps of Engineers. Coastal Engineering Research Center, Flort Belvoir, VA, 110 pp.

Garbisch, E.W., 1986. Highways and Wetlands: Compensating Wetland Losses. US DOT, FHWA Report No. FHWA-IP-86-22 Washington, DC, 60 pp.

Garbisch, E.W., 1989. Wetland enhancement, restoration, and construction. In: Majumday, S.K., Brooks, R.P., Brenner, F.J., Tiner, R.W. (Eds.), Wetlands Ecology and Conservation: Emphasis in Pennsylvania. The Pennsylvania Academy of Science, Philadelphia, pp. 261–275.

Garbisch, E.W., 2002. The Dos and Don'ts of Wetland Construction (Creation, Restoration, and Enhancement). Environmental Concern Inc., St. Michaels, MD, 180 pp.

Horton, T., 2003. Turning the Tide: saving the Chesapeake Bay, second ed. rev. and expanded. Island Press, Washington DC.

Sipple, W.S., 1990. Exploring Maryland's freshwater tidal wetlands. Atlantic Nat. 40, 3–18.

Tiner, R.W., D.G. Burke, 1995. Wetlands of Maryland. U.S. Fish and Wildlife Service, Ecological Services, Region 5, Hadley, MA and Maryland Department of Natural Resources, Annapolis, MD. Cooperative Publications, 193 pp. (plus appendix).

Available online at www.sciencedirect.com

SCIENCE DIRECT°

Ecological Engineering 24 (2005) 309–329

ECOLOGICAL ENGINEERING

www.elsevier.com/locate/ecoleng

ELSEVIER

Landscape restoration following phosphate mining: 30 years of co-evolution of science, industry and regulation

Mark T. Brown *

Department of Environmental Engineering Sciences, University of Florida, Gainesville, FL 32611, USA

Received 24 November 2004; received in revised form 28 January 2005; accepted 28 January 2005

Abstract

The restoration of phosphate mined lands in Florida is large scale, potentially covering over 300,000 acres (121,000 ha), and rivals other restoration efforts like the Florida Everglades in size and complexity. The issues surrounding mining and subsequent restoration of the landscape are global, national, and local in scale. The entire system of phosphate mining and restoration involves local citizens, governmental agencies, research scientists, and industry personnel in a program that might be seen as adaptive management. It is suggested that restoration is managing adaptive self-organization of the ecosystems and landscapes and that it is the domain of ecological engineering. The past 30 years of research concerning various aspects of landscape restoration after phosphate mining are elucidated, and the research's relationship to management and regulation are discussed. Finally, the complex issues that are inherent in large restoration programs are discussed and it is suggested that a cooperative environment and vision may be the key elements that are missing.
© 2005 Published by Elsevier B.V.

Keywords: Landscape; Phosphate mining; Ecosystems

1. Introduction

Phosphate mining in Florida is open pit mining (see Fig. 1) where an overburden of approximately 10 m is stripped off and set aside and the clay/sand calcium phosphate matrix is removed.[1] During beneficiation, the calcium phosphate is separated from the sand and clay. The clays, now in a slurry mixture, are returned to

the lands and stored in elevated clay settling areas that occupy approximately 40% of the post-mining landscape. The majority of research on phosphate mine restoration, since it began in the mid 1970s, has been concerned with the restoration of the remaining mined lands that are not used for waste clay storage. Very recently, research has been initiated to evaluate the potential of clay settling areas as the setting for wetlands creation.

Landscape restoration of phosphate mined lands has benefited at times from a tight coupling of research, applied ecological engineering and government regulation. This coupling resembles co-evolution (or recip-

---

* Tel.: +1 352 392 2309.

  E-mail address: mtb@ufl.edu.

[1] Definitions of terms used in this manuscript are given in Appendix A.

0925-8574/$ – see front matter © 2005 Published by Elsevier B.V.
doi:10.1016/j.ecoleng.2005.01.014

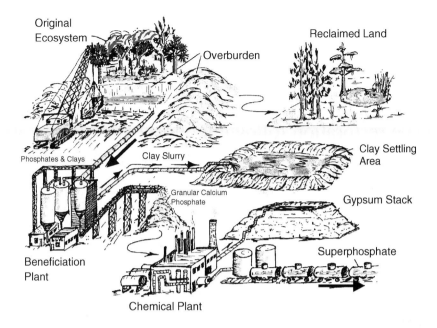

Fig. 1. Schematic diagram of phosphate mining in Florida. The phosphate matrix, at depths of 6–10 m below the ground surface, is open-pit mined where the soils on top of the matrix (overburden) are piled to the side in spoil piles and the phosphate matrix, high in clay content, is slurred and pumped to the beneficiation plant. The by-products from beneficiation are: clays, which are pumped into elevated clay settling ponds, and sand-tailings used to back-fill mined areas (not shown in the illustration). Granular calcium phosphate is converted to super phosphate fertilizer in chemical plants producing a gypsum by-product that is stacked high near the plant. The final land uses after mining are reclaimed land (about 50–60% of the landscape), clay settling areas (40% of landscape) as well as chemical plants, transportation and gypsum stacks (about 10%) which most frequently are constructed on unmined land.

rocal evolutionary change) of: (1) the research community as the need for answers to the question "how do we restore the landscape?" increased; (2) industry, as they applied new techniques as fast as they were being researched and documented; and (3) government regulation of restoration, as agency personnel learned what industry was capable of and what researchers suggested was possible. In some respects, this co-evolution resembles what has been referred to as "adaptive management" (Holling, 1978; Walters, 1986; Walters and Holling, 1990) or the systematic process for continually improving management policies and practices by learning from the outcomes of operational programs. Taken as a whole, the research community mostly focused in the academic world, industry, represented by numerous companies and government, including local, state and federal regulatory agencies represents a relatively complex system of actors. Nonetheless, all have as a common goal the restoration of lands in Florida that have been mined for phosphate ore.

In this paper, the 30-years experiment in landscape restoration following passage of legislation by the State of Florida requiring reclamation of phosphate mined lands is described. Much of the research that has been conducted during that time is summarized and related to the ecological engineering practices of industry and changes in the regulatory program of State of Florida's Bureau of Mine Reclamation (BMR).

1.1. Brief historical perspective

Phosphate mining in Florida began in the late 1800s with hundreds of small hard rock mines in north and central Florida. About the same time, pebble phosphate was discovered in and around the Peace River in south/central Florida. This region, known as the Bone Valley, soon dominated phosphate production because of the relatively lower costs of production for pebble phosphate compared to the hard rock. As a result, most mining for hard rock through out the rest of the state dwindled and ceased by the early part of the 1900s.

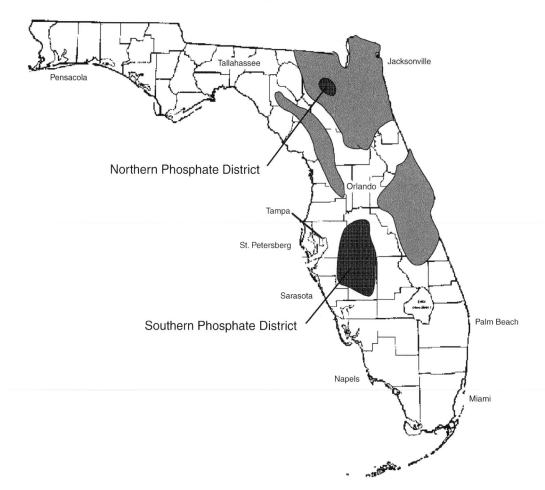

Fig. 2. Map of the phosphate regions of Florida. Gray areas are districts of secondary "reserves", while the dark areas are the primary phosphate districts. The southern most primary district is also known as the "Bone Valley" formation.

North Florida phosphate mining began in the mid 1960s and continues today near White Springs, FL. Currently there are two active mining areas in Florida (Fig. 2) known as the northern and southern phosphate districts, where about 5000 acres (2000 ha) are mined each year.

The area dominated by phosphate mining totals about 300,000 acres (121,400 ha). In 1990, there were 11 phosphate companies operating in Florida; by 2004, as a result of the changes in ownership and corporate buyouts, there were 3. Each company may have numerous mines. Phosphate mines range in size from about 4500 acres (1800 ha) to about 21,000 acres (8500 ha) with the average size of about 10,000 acres (4000 ha). Each mine is usually planned, mined and reclaimed as a single entity. In the early years, little attention was

given to the lands outside the mining unit but increasingly, regulatory agencies are requiring a broader perspective from industry as mines are planned and eventually reclaimed. Today each mine is permitted by the state and part of the permit application is a conceptual reclamation plan that includes all the lands and details regarding their restoration. The BMR reviews the plans for consistency with adjacent mines and land uses, and with its regional conceptual plan.

Presented next is a review of the State's regulation of mined land restoration. It is important at this point to call attention to the difference between restoration and reclamation. Restoration (in this paper ecological restoration) is the process of assisting the re-establishment of natural communities, habitats,

species populations or other ecological attributes that have been eliminated or greatly reduced on a given location. Reclamation is a more general term and means the process by which lands disturbed, as a result of mining activity, are reclaimed back to a beneficial land use. A subtle difference, but important. Reclamation as used here only refers to returning lands to a beneficial use, while restoration implies restoring ecological and hydrological functions.

1.2. State regulation of reclamation

Throughout the many years of phosphate mining in Florida, there was no requirement for reclamation or restoration of mined lands. The industry only reclaimed lands where there was an economic incentive or an overwhelming aesthetic need. In 1975, the State of Florida passed legislation that mandated all land mined for phosphate after July 1, 1975 must be reclaimed. Further, the State provided some funds from the severance tax on mined phosphate to assist the industry in the reclamation of lands mined prior to 1975. The goal, of course, was to reclaim all lands including those mined prior to the enactment of the reclamation law.

Beginning in 1975, the Florida Department of Natural Resources (FDNR) was assigned the task of regulating reclamation. In response, FDNR adopted rules that covered reclamation (Chapter 16C-16, Florida Administrative Code (FAC)). At first, the rules were designed to hide the evidence of mining. During the first years of regulation, from 1975 to about 1980, phosphate mining companies were required to level spoil piles and plant 10% of the lands in trees. The resulting landscape of recontoured mine pits and uplands was called land and lakes. During the early 1980s, State regulation began to evolve, requiring reclamation of wetlands, according to the quantitative success criteria that required survival of 400 trees per acre, 80% cover by desirable species and no visible evidence of erosion. Beginning in the mid 1980s, State regulations became more prescriptive with several revisions of the States Reclamation Rules. Appendix B is a summary of current reclamation rules related to the landscape restoration after phosphate mining.

It should be noted that prior to the passage of legislation and rule making that resulted in Chapter 16C-16 FAC, requiring the phosphate industry to reclaim the lands they mined, there was little or no incentive to do

so. Therefore, there was little or no restoration research conducted by industry or the research establishment. The FDNR rules encouraged the phosphate industry to begin restoration research in earnest in the early 1980s. This is not to say that industry is to blame or that they were insensitive to the ecological concerns, it is only to demonstrate that regulation, research and implementation are all necessary to achieve the goals of ecological restoration of drastically altered lands. Regulation has a tendency to direct industry to seek alternatives and research can guide those alternatives.

1.3. State of Florida sponsored restoration research

In addition to the legislation that required reclamation, the State of Florida recognized the need for research and passed legislation that funded phosphate research through a statewide research institute. In 1978, the Florida Legislature created the Florida Institute of Phosphate Research (FIPR) by (Chapter 378.101, Florida Statutes) empowering it to initiate, conduct and sponsor the studies to minimize or rectify any negative impact of phosphate mining and processing on the environment and improve the industry's positive impact on the economy. This includes developing better techniques for reclaiming land and developing more efficient mining and processing technologies. The Institute is financed with funds from the state severance tax on phosphate rock. Over the years, FIPR has funded most of the academic research related to the restoration of phosphate mined lands. Its influence on the direction and overall output of restoration research cannot be over-emphasized. Through its research priorities set by technical advisory committees, made up of stakeholders from government, industry, environmental groups and academics, it has had a major impact on the restoration of mined lands. Table 1 is a partial list of the restoration research projects funded by FIPR.

2. Phosphate mining in perspective

2.1. Phosphate contribution to Florida, USA and the World

Table 2 lists some relevant facts regarding phosphate mining in Florida. It is quite apparent that phos-

Table 1
Partial list of restoration research projects funded by the Florida Institute of Phosphate Research

Project title/agency	Date funded
Development of techniques for the use of trees in the reclamation of phosphate lands, Division of Forestry	06/13/1980
Enhanced ecological succession following phosphate mining, University of Florida	03/20/1981
Interactions between phosphate industry and wetlands, University of Florida	03/20/1981
Ecological considerations in lake reclamation design for Florida's phosphate region, Environ. Science & Engineering Inc.	10/16/1981
Study of a method of wetland reconstruction after phosphate mining, University of Florida	08/20/1982
Pelletization of seeds, University of South Florida	09/15/1983
Development of techniques for reclamation of mined land, University of Florida	01/01/1984
Measurement of recovery in lakes following phosphate mining, VPI & SU	03/01/1984
Interactions of wetlands with phosphate mining, University of Florida	03/01/1984
Water quality in lakes in Central Florida phosphate region, Post, Buckley, Schuh & Jernigan Inc.	05/01/1984
Vegetable production potential of sand–clay mixes, Bromwell & Carrier Inc.	06/18/1984
Propagation and establishment of indigenous Florida plants for revegetation & restoration of phosphate mining sites, University of Florida	10/01/1984
Viability of wetland topsoil for reclamation, FIPR (In-House)	08/01/1985
Restoration techniques for sand scrub, Florida Southern College	01/17/1986
Citrus groves on reclaimed mined lands, Zellars-Williams Co.	06/17/1986
Production of high cash value crops on mixtures of sand tailings and waste phosphatic clays, Bromwell & Carrier	12/20/1986
Hydrologic impacts of phosphate mining in small basins, USGS	04/01/1987
Alternatives for restoration of soil, University of Florida	05/01/1987
Mined Lands agricultural demonstration project, Polk Co. BOCC	03/01/1988
Reintroduction of gopher tortoises to reclaimed, L. Macdonald	05/09/1988
Enhancing tree revegetation on phosphate surface-mined land, FIPR/Clewell	08/29/1988
An evaluation of benthic meiofauna and macrofauna as success criteria for reclaimed wetlands, University of Florida	06/13/1989
Improving upland native plant revegetation potential, USDA Soil Conservation	11/13/1991
Equipment for construction of drainage systems on clay settling areas, Polk County BOCC	03/01/1992
Macro & meiofaunal distributions in headwater streams of the alafia river, Florida, University of South Florida	01/01/1993
Studies of wildlife usage and restoration of upland habitats on phosphate mined land in Central Florida, University of South Florida	04/16/1993
An evaluation of constructed wetlands on phosphate mined lands in florida: vegetation, soils, aquatic fauna, water quality, ecosystem analysis, and values, functions & regulations, University of Florida	06/25/1993
Ecology, physiology and management of cogongrass (imperata cylindrica), University of Florida	11/15/1993
Plant competition effects on tree establishment and growth on phosphate mined lands, FIPR (In-House)	03/01/1994
Hydrology and water quality of reclaimed phosphate clay settling areas in West-Central Florida, U.S. Geological Survey	02/01/1995
Evaluation of feasibility of water storage reservoirs on mined lands to meet future agricultural, industrial and public water supply demands, Schreuder Inc.	07/01/1995
Wildlife usage of mesic flatlands and its bearing on restoration of phosphate mined land in Central Florida, University of South Florida	08/11/1995
Managing runoff water quality from clay settling areas used for intensive agricultural production, University of Florida	09/01/1995
Shading effects on nuisance species and succession on phosphate mined lands, University of Florida	03/01/1996
Post-mine reclamation of upland communities, Jones, Edmunds & Associates Inc.	12/01/1996
Development of seed sources and establishment methods for native upland reclamation, USDA	01/02/1997
Synthetic seed production of florida's indigenous plants, University of Florida	04/15/1997
Managing weed competition and establishing native plant communities on reclaimed phosphate mined lands in Florida, FIPR (In-House)	09/01/1998
Self-organization and successional trajectory of constructed wetlands on phosphate mined lands in Central Florida, University of Florida	09/01/1998
Water quality investigation of in situ tailing-sand deposits under natural environmental conditions, Schreuder Inc.	09/01/1998
Rapid production of Florida's indigenous plants via micropropagation, University of Florida	02/19/2000

Table 2
Phosphate mining facts (Florida Phosphate Council, 2004)

Mined area to date[a]	130,000 ha
Mining rate	2–2500 ha/year
Total to be mined	280,000 ha
Phosphate rock extracted (2003)	28.7 million MT
Taxes paid (2003) (in US$)	83 million/year
Annual operating expenses (2003)	
Yearly wages (in US$)	445 million/year
Equip and supplies (in US$)	1360 million
Electricity (in US$)	127 million
Services (in US$)	165 million
Investment in facilities (total) (in US$)[a]	10 billion

[a] Florida Institute of Phosphate Research (2004), http://www.fipr.state.fl.us/southintro.htm.

phate mining is an important part of Florida's economy, an essential part of USA agriculture and when placed in a global perspective, a critical component of the world food system. Florida contributes nearly 75% of the USA's phosphate needs and about 25% of the global needs. In 2003 (the last year for which there are complete data (Florida Phosphate Council, 2004)), the phosphate industry extracted 28.7 million metric tonnes of phosphate rock from 4501 acres (1800 ha) of land in Florida. Its contribution to the economy of Florida may be as much as US$ 10 billion in 2003 from the phosphate and related chemical industries in an economy that totaled about US$ 500 billion. The economic value of phosphate rock in Florida, based on a price of US$ 28 per metric tonne, was about US$ 800 million in 2003.

2.2. Emergy evaluation of phosphate and restoration

The monetary value of phosphate rock, based on its current price on the global market, provides a benchmark for what its value is to the market economy. The market price, however, does not provide an estimate of its value as a raw resource used to drive agricultural productivity, instead it provides an estimate of its utility value based on willingness to pay. Another way of estimating its value at the larger scale is related to the total energy required to make it (or its emergy). To do this, it is necessary to determine all the energy from natural processes required for its formation.

There are several theories related to the formation of Florida's phosphate deposits. The most plausible is illustrated in the left side of Fig. 3. The actions of acid waters, formed as a result of organic matter decomposition, carry phosphorus from terrestrial vegetation where it precipitates with calcium carbonate to form calcium phosphate in formations underlying Florida. Odum (1996) estimated the emergy required to concentrate phosphate in this way to be 3.9 E9 sej/g. Mining of phosphate shown on the right side of Fig. 3 requires significant emergy in materials, fuels and human service (5.4 E18 sej/ha). The emergy yield, shown leaving the diagram to the main economy on the right, is over three orders of magnitude greater than the purchased inputs (6.1 E22 sej/ha). When the current price for phosphate rock (about US$ 28 /MT) is expressed in emergy, using a emergy per dollar ratio of 1 E12 sej/$, the emergy equivalent of its market price is about 2.8 E13 sej/MT or 2.8 E4 sej/g. When compared to emergy of the phosphate rock itself, it is quite apparent that the difference between the value received in the phosphate rock and the emergy represented by the price paid is extremely high. The value received is about five orders of magnitude greater than the price paid (3.9 E9 sej/g/2.8 E4 sej/g = 1.4 E5 to 1). This difference suggests the extremely high value one currently receives when purchasing phosphate rock.

2.3. Net benefits of restoration

Ecological restoration, to be successful, should restore ecosystem function. The function and structure of the ecosystems are defined by two aspects: (1) abiotic conditions (temperature, soil type, terrain, disturbance regime, etc.) and resources (water, nutrients, sunlight) and (2) the trophic structure or feeding relationships among the species that shapes the flow of energy through the ecosystem. Ecosystem disruption results in a breakdown of both the function and structure. An impact may disrupt abiotic components (soil structure might be changed, as in mining) as well as trophic structure (the loss of primary producers, as when land is cleared). Restoration efforts should be focused on reestablishing pre-disturbance ecosystem functions and to be effective, it should not cost more than it yields.

The graph in Fig. 4 illustrates the effect of disruption on the ecosystem function. The disordering influence (in this case, phosphate strip mining) impacts ecosys-

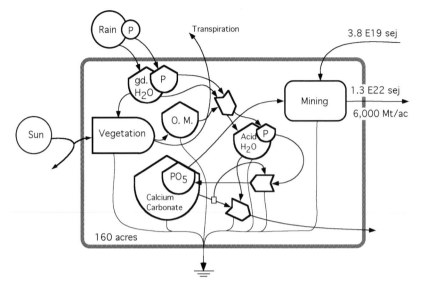

Fig. 3. Systems diagram of phosphate formation and mining. The time frame is 1800 years for formation of phosphate rock and a matter of months for its mining. The flows of emergy for mining and emergy of the yield are for mine life.

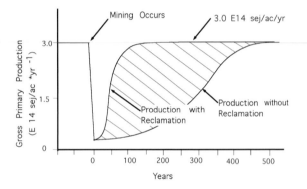

Fig. 4. Natural restoration of mined lands in Florida may take as long as 500 years because of the lack of near by seed sources, while ecologically engineered restoration may only take a little over 100 years (Weber, 1994). The emergy benefit of restoration can be estimated as the area between the two graphs and related to the emergy costs as a benefit cost ratio. It is suggested that effective restoration should have a ratio greater than 1.0.

tem function causing it to decline to very low levels in a short period of time. Left to its own devices, the landscape would re-establish ecological productivity and eventually ecosystem function through natural restoration processes of landscape and ecological succession. Because of the areal extent of phosphate mining and distance to seed sources, natural restoration is estimated to take about 500 years (Weber, 1994). Restoration can be sped up or enhanced, however, through the ecological interventions, for instance, the introduction of seeds and planted vegetation. The difference between the two lines on the graph (the area between the two lines) is the benefit from restoration.

Fig. 5 illustrates that landscape restoration after phosphate mining in Florida consists of recontouring spoils, planting trees and sowing seeds to stabilize soils. An additional input is the human services in engineering design and environmental monitoring. Emergy costs for a typical restoration project are shown in Fig. 5. The inputs are for a 160 acre (65 ha) restoration project in central Florida and assumes 100 years of renewable inputs to reach full replacement of ecological functions.

Quantitatively, the net benefit of phosphate restoration can be estimated if it is assumed that ecosystem function is equal to emergy of gross primary production (GPP). Using Fig. 4 as a guide and the data from Fig. 5, the net benefit from restoration for a typical 160 acre (65 ha) mine can be calculated. Brown and Bardi (2001) estimated emergy value of GPP for average Florida ecosystems as 3.0 E14 sej/(acre year) (7.5 E14 sej/(ha year)). Using the hypothetical graphs in Fig. 4, the benefit from restoration of 160 acres (the area between the two lines = 6.6 E16 sej/acre) is 10.6 E18 sej (6.6 E16 sej/acre × 160 acres = 10.6 E18 sej).

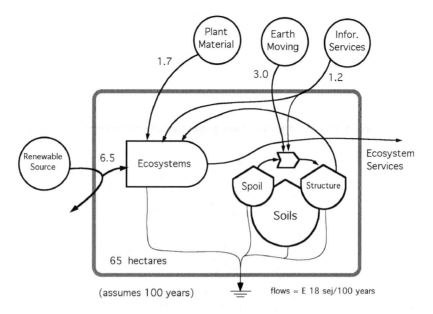

Fig. 5. Total emergy required to restore 65 ha of the Florida landscape following phosphate mining, assuming 100 years to restore ecological functions. Values are total emergy over 100 year period, calculated as follows:

Renewable inputs equal to water use in transpiration
    transpiration = 0.77 m/year
    Gibbs free energy of water = 4.94 J/g
    energy in water used = (0.77 m) (1.00E + 04 m^2/ha) (1.00E + 06 g/m^3) (4.94 J/g)
    = 3.80E + 10 J/ha/year × 100 years × 65 ha = 2.5 E14 J
    transformity = 26,096 sej/J (calculated as weighted average of rain and run-in)
Plant material (planting of seedlings and sowing seed)
    seed 11.4 kg/ha × 65 ha × 4.7 E12 sej/kg seed = 3.5 E15 sej
    planted 10 species at 1 E16 sej/spp = 1 E17 sej
    fuel use = 1.8 E7 J/ha × 65 ha × 6.6 E4 sej/J = 7.7 E13 sej
    service in planting = $24,600/ha × 65 ha × 1 E12 sej/$ = 1.6 E18 sej
    total = 1.6 E18 sej + 1 E17 sej + 7.7 E13 sej + 3.5 E15 sej = 1.7 E18 sej
Earth moving
    fuel use = 3.2 E8 J/ha × 65 ha × 6.6 E4 sej/J = 1.4 E15 sej
    service in earth moving = $4.6 E4 /ha × 65 ha × 1 E12 sej/$ = 3.0 E18 sej
Information services in engineering design and monitoring
    services = $18.4 E3/ha × 65 ha × 1 E12 sej/$ = 1.2 E18 sej

The emergy costs of the restoration (sum of inputs from Fig. 5) is 5.9 E18 sej and the emergy benefit cost ratio is about 1.8:1.

3. Phosphate restoration research

Most restoration research related to phosphate mined lands in the early years was primarily concerned with wetlands. There were several reasons for this, the most important of which was the fact that ap-proximately 15% of the Bone Valley district of Polk and eastern Hillsborough and Manatee counties is overlained by wetlands. Native wetlands include both the herbaceous ecosystems (wet prairies and marshes) and forested or shrub-dominated ecosystems (cypress domes, cypress swamps, mixed cypress–hardwood swamps, bay forests and swamp thickets). In addition to the large wetland acreage, the phosphate industry has been required by law to reclaim wetlands since the adoption of FDNR's rules, which stated that the wet-lands affected by mining operations were to be restored

to at least pre-mining surface areas. In other words, wetlands had to be restored acre for acre and type-for-type.

Restoration begins with the recontouring of spoils and the creation of a landscape that includes areas of low topographic relief where surface and groundwaters will contribute to an appropriate hydrologic regime, which will support hydrophytic vegetation. Two techniques have been widely used to introduce wetland vegetation to created wetland areas: mulching and handplanting. Mulching is the practice of applying organic soils, obtained from wetlands that will be mined, to restoration sites. The thickness varies from a few centimeters to tens of centimeters. These organic soils contain seed material, rhizomes and other genetic material to jump-start restoration. Both planting of the desirable species and mulching are often applied on a given site to further enhance the successful establishment of vegetation.

3.1. Early research focused on wetland restoration

The earliest restoration research was carried out by the phosphate companies as demonstration projects often as part of their permitting objectives. In the 5 years following adoption of 16C-16 FAC, most companies had some type of demonstration project, the majority of which were focused on evaluating the potential for herbaceous wetland restoration. Robertson (1985) provides a detailed summary of these early trails by the industry. Several of these industry-sponsored projects evaluated the technique of mulching in the newly constructed wetlands.

In an early test of mulching, Brown et al. (1985) applied mulch in a block design in varying thicknesses. The mulch contained large numbers of propagules and seeds from which a diverse vegetative community developed (Fig. 6). Water levels in the experimental site were lower than the initial design called for, but nonetheless, a full coverage of wetland species germinated from the seedbank.

Most of the early wetland restoration research concentrated on the techniques for establishing a diverse vegetative cover as defined by the Rules of FDNR. An exception was the research by Odum et al. (1983) documenting the interactions between the phosphate industry and wetlands. This multifaceted study evaluated landforms created by mining, the use of detritus as a seed source, tree planting on waste clay sites and the

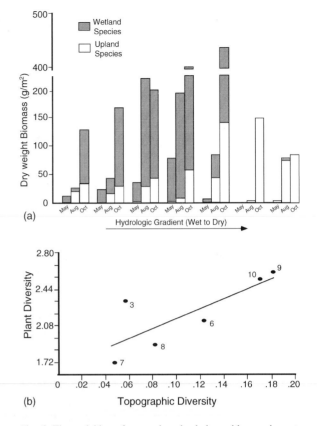

Fig. 6. The mulching of restored wetland sites with organic matter from existing wetlands introduces seeds and rhizomes and other genetic material to jump-start ecosystem development. Hydrology is also an important variable. In the top graph, total biomass and wetland species increase along the hydrologic gradient from wet to dry until they decrease and upland species dominate. An important added benefit of mulching is the topographic diversity that results from uneven application and which translates into higher species diversity because of the variation in "hydrologic habitats" (Brown et al., 1985).

use of tree cores from cypress trees to monitor stress in wetlands. In a continuation of that project, Rushton (1988) focused on the wetland establishment on waste clay sites and documented several general trends that not only applied to waste clay sites but also the overall phosphate mined landscape. These were as follows:

- A pattern of arrested succession emerged that seemed to be related to the distance to seed source.
- The soils characteristic of mined lands tend to have higher concentrations of clay than native soils. These clayey soils tend to lower air and water movement and can range from quite sticky when wet to brick-

hard when dry. They also exhibit considerable expansion and contraction depending on the moisture regime. Soil development is a most important issue.

• Soils and water in phosphate mined areas are high in phosphorus and tend to naturally favor low diversity ecosystem types that are characterized by early successional colonizing species.

3.2. The focus of restoration research shifts to landscape scale

By the early 1980s, it became apparent that there were critical research needs related to landscape scale restoration. The cumulative area affected by phosphate mining by this time totaled about 200,000 acres (81,000 ha) and the requirements for reclamation had been in effect for over 5 years. Until about 1983, little or no attention had been paid to larger scale restoration issues but instead, research had focused primarily on wetlands and to a lesser extent upland ecosystem restoration. In 1984, the FIPR funded a 5-year restoration research project that was focused on integrated landscape restoration. The study's goals were to develop guidelines for landscape planning and design through studies of Florida watersheds and ecosystems and the integration across landscapes of wetlands, forests and agricultural uses after restoration. Titled "Techniques and Guidelines for Reclamation of Phosphate Mined Lands as Diverse Landscapes and Complete Hydrologic Units" (Brown and Tighe, 1991), the project was designed to provide the phosphate industry with information about the structure and organization of watersheds (sizes, slopes, upland/wetland ratios, spatial organization, etc.) and the structure and organization of ecosystems (dominant species in all strata, soils, topography, hydrologic regime, etc.). In addition, the project team also studied mined lands to characterize typical abiotic conditions in order to cross reference mined lands with appropriate ecosystem types. The results of the study, viewed as a restoration cookbook by its authors, included: (1) design principles and metrics for watersheds, (2) characteristics and design guidelines for streams and floodplains, (3) vegetation and structural characteristics of native ecosystems, and (4) hydrologic regime of native ecosystems.

The study paved the way for additional large-scale research when it concluded that research was needed to study long-term trends of reclaimed landscapes and suggested six areas for improvement and future research:

1. Restoration of upland and wetland forest ecosystems concentrated exclusively on planting tree species and under-story vegetation was completely absent from restoration planning. A shift in restoration philosophy was needed to insure that species in other strata were included in the restored forested communities.

2. As the lands are reclaimed and surrounding lands are developed, surface hydrology and the overall hydrologic function of the Peace River should be of paramount concern. A regional planning effort is needed to define new watershed boundaries and begin to propose reclamation schemes that enhance regional hydrology.

3. In many areas of the phosphate district, especially older restored lands, non-native vegetation has become important components of the plant community. The long-term consequences of and role of the exotic species on restored lands will become more and more important, especially as mining continues to move southward.

4. The use of phosphate-mined lands for the recycle of treated sewage effluent and sewage sludge should be investigated.

5. The continued practice of requiring the elimination of cattails, primrose willow and Carolina willow (often termed undesirable species) from restoration sites should be re-evaluated.

6. The role of "seed islands" to enhance natural restoration of mined lands through wind and animal seed dispersal should be investigated.

In the late 1980s, recognizing the need for an integrated effort to plan the restoration of the phosphate district, the FDNR's Bureau of Mine Reclamation (BMR) began an Integrated Habitat Network (IHN) plan and published "A Regional Conceptual Reclamation Plan for the Southern Phosphate District" in 1992 (Cates, 1992). The BMR designed the IHN plan (shown in Fig. 7) to be a guide for the reclamation of mined phosphate lands throughout the southern phosphate district. The core lands of the habitat network were the riverine floodplains along major rivers of the district, which were already State property. Lands adjacent to these core areas were to be reclaimed and act as buffer lands

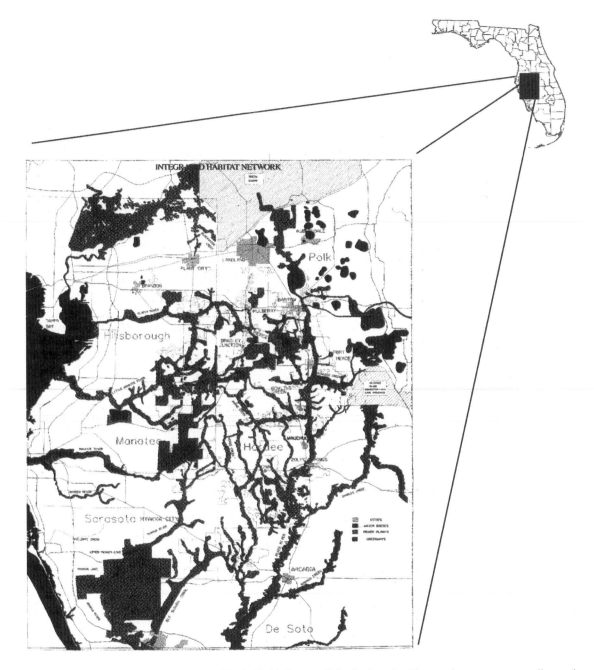

Fig. 7. Integrated Habitat Network (IHN) as proposed by the Florida Bureau of Mine Reclamation. The area shown encompasses all or portions of nine counties totaling 2.56 million acres (1.44 million ha). By 2025, most of the central portion of the map will have been mined. The concept for the IHN is to have a regional network whose core is composed of floodplain ecosystems of streams and rivers connecting several state parks and reserves (rectangular areas). "Adjacent to the floodplain will be a zone which contains mesic/transitional forests, upland forests and other habitats considered "critical" and in need of protection. In general, the progression will be from less-intensive land uses near the floodplain to more-intensive as distance from the floodplain increases" (Cates, 1992).

that would compliment and enhance the habitat value of the core lands, benefit the water quality and quantity in the area and serve as upland habitat connections between the mining region's rivers and significant environmental features outside the mining region (Cates, 1992).

In June 1988, FIPR funded the Central Florida Regional Planning Council to develop a geographical database for the southern mining district. The database was to depict present land use and projected land use to 2010. Recognizing that by 2010, most of the minable phosphate ore will have been extracted from the northern portions of the district and the industry's emphasis will shift almost exclusively to restoration, the study's goals were to develop a GIS database of land use and land cover with projections to 2010. The issues that fostered concern and lead to the regional planning effort were related to the industry's utilization or disposition of their land areas which were being reclaimed and released by the permitting agencies. Some firms were expected to retain ownership of their reclaimed land and use them for agricultural, timber or cattle production. In other areas, because of their close proximity to expanding urban areas, there was the potential for industrial land uses such as waste treatment facilities, power plants, warehousing and other enterprises which could take advantage of the low residential density and in-place infrastructure (roads, railroads, power lines, deep wells, etc.). The regional area was also experiencing population growth pressures and an increased demand for residential, commercial and recreational land. It was felt that the phosphate industry was in a unique position to contribute to the orderly and environmentally sound development of the region (Long and Orne, 1990).

3.3. Restoration research becomes regulation driven

Following the 1980s era of macro-scale restoration research, there was a shift towards research that addressed some of the "regulatory dogmas" that were driving restoration. Chief among these was that a successfully restored ecosystem should not contain what state and federal agencies called undesirable species. These included cattail (*Typha* spp.), primrose willow (*Ludwigia peruviana*) and Carolina Willow (*Salix caroliniana*), among others. Several studies (Brown et al.,

1998, 2000; Richardson and Kluson, 2000) were conducted with funding from FIPR to evaluate the effects of so-called nuisance species at the ecosystem level as well as on the growth and survival of planted trees on the restored sites. Simulation models were used to predict the effects of nuisance species on the ecosystem function showing that while they may dominate during early ecosystem development, they soon declined in abundance and importance, but in so doing provided an important service by increasing soil organic matter and nutrient retention (Fig. 8). Field evaluations and simulation models of competition, and structural and functional analysis of ecosystems with and without these species concluded that pioneer species were not as detrimental to community development and in fact may facilitate development.

This was a time of reflection as well. To take stock and learn from restoration efforts to date, FIPR funded several projects that took hard looks at restoration over the 20-year period since the beginning of restoration in 1975. A coalition of researchers, in 1995, from private consulting, environmental organizations and universities embarked on a 2-year study of wetlands restoration by the phosphate industry (Erwin et al., 1997). The primary research task of this ambitious project was to assess and analyze the available database for constructed wetlands on phosphate mined lands in Florida and, where necessary, to supplement existing data with limited additional sampling and computer modeling. Research goals were directed at determining current technical and operational success of created wetlands to develop as persistent, functioning and integrated ecosystems. This was accomplished through an evaluation of design criteria and the wetland structure and function that has developed on the existing sites. The project team identified six specific research goals as follows:

1. to provide a database from existing studies to guide operational and policy changes needed to improve design, construction, monitoring and assessment of the constructed wetland projects, and to determine the adequacy of the existing database, providing recommendations to ensure the utility of future research and monitoring efforts;
2. to determine the extent to which existing constructed wetlands are persistent, functioning ecosystems;

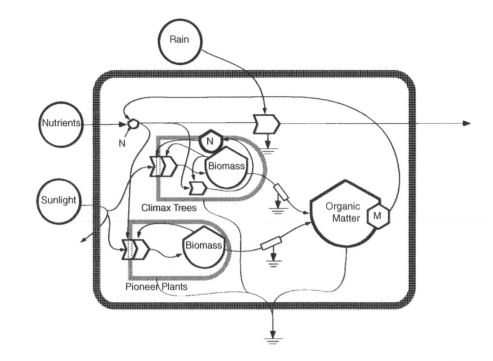

(a)

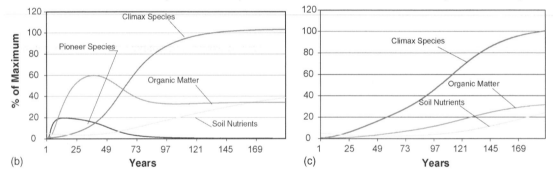

(b)   (c)

Fig. 8. Simulation model of role of early successional pioneer species in ecosystem development. (a) The competition for sunlight and nutrients between pioneer species and climax species. (b) Graphs of simulation results show ecosystem development is facilitated by pioneer species, primarily in the buildup of organic matter and nutrients. (c) Lower overall storage of organic matter and delayed development of mature community result when pioneer species are eliminated.

3. to determine whether constructed wetlands are properly located in the reclaimed landscapes;
4. to determine ecosystem functions and values provided by the constructed wetlands, to identify appropriate indicators of functions and values, and to develop quantitative methods of measuring those indicators;
5. to determine how success criteria should be applied in evaluating the attainment of goals and of devel-

opment trends for constructed wetland projects;
6. to identify future research needs of industry and regulatory agencies.

The study concluded with four main areas of concern:

- The overall adequacy of data was poor, limited or lacking in standardization to do quantitative evaluations of constructed wetland success.

- Many of the constructed wetlands observed were apparently persistent. Constructed wetlands were providing similar functions as natural wetlands but at different capacities. Most constructed wetlands provide wildlife functions but sometimes for different groups of species than typically found in similar undisturbed wetlands.
- Reclaimed mine lands are disconnected and dominated by agriculture (primarily pasture and grazing lands) with numerous fragmented habitats and watersheds. Since the reclaimed landscape is often a patchwork of reclamation projects in various stages of design, implementation and successional regrowth, it continues to be a challenge to link reclamation projects and their natural ecological communities together in a cohesive regional habitat network. However, current approaches to reclamation and reclamation planning have improved these linkages to provide a habitat network.
- Reclamation goals should establish a landscape plan for an entire watershed; types and sizes of habitats, hydrologic pathways, topography, types and levels of functions to be provided. Success criteria should be a measurable criteria used to assess the degree of goal attainment. As built surveys of wetlands and topography, post-construction aerial photographs can be used to document the size and configuration of landscape features.

The research team suggested over a dozen needed research topics that spanned scales from regional planning issues to consequences of herbicide spraying in newly constructed wetlands. The document described in detail the successes and failures of past restoration efforts and indicated where industry and the regulatory agencies needed change. Unfortunately, to date it appears that the recommendations have been largely ignored.

3.4. Current restoration research

As the phosphate ores of the northern portions of the southern phosphate district are mined out, the industry is moving southward. The permitting of new mines is a lengthy process and often contentious. There appears to be significant local opposition to expanding the mined area farther south, and stakeholders have used several environmental issues as key elements in their challenges. Chief among these is the long-term restoration of these newly mined lands into a functional landscape. There is particular concern regarding the area of land that will be dominated by waste clay settling basins, estimated to occupy as much as 40–60% of the landscape after mining. Another is the effect of mining on regional hydrology. Current research funded by FIPR is addressing both of these topics as well as wildlife utilization of restored lands.

What may be needed is funding of planning research that provides a vision of how the landscape will look and function after mining has ceased and restoration is complete. Without such a vision, there is not a clear goal in mind for measuring restoration success. For instance, the landscape illustrated in Fig. 9 is a conceptual restoration plan for a mine that shows agriculture and wildlands that are integrated into the larger landscape. Hydrologic connections are also stressed. The mosaic of upland and wetland forests that dominates the western portions of the restored landscape is a wildlife corridor that creates a link across the landscape. Large-scale restoration may require an integrated effort of citizens, industry, government and planning agencies to articulate a vision for the entire region and may be an essential exercise to develop consensus and restoration goals.

4. Restoration of phosphate mined lands: an adaptive endeavor

Adaptive is defined as taking available information into account or able to be adjusted for use in different conditions. One way of looking at this, when applied to living systems, is related to what has been termed adaptive self-organization (Laszlo, 1972) or the ability for living systems to use new information. A second way might refer to the act of management as in adaptive management (Gunderson et al., 1995; Holling, 1978).

4.1. Ecological engineering and adaptive self-organizations

The management of nature's adaptive self-organization is the field of ecological engineering. With the increased rates and spatial scales of human induced

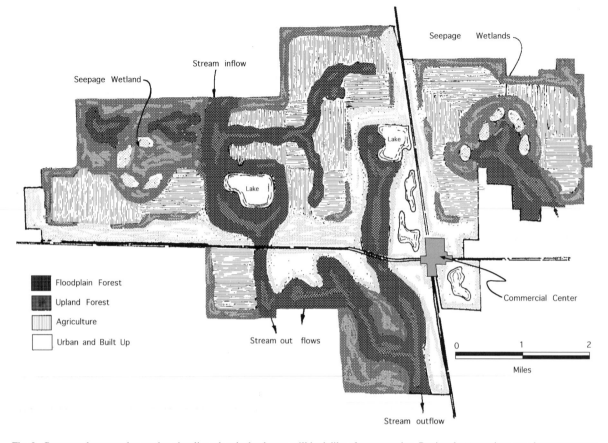

Fig. 9. Conceptual master plan used to visualize what the landscape will look like after restoration. Regional master plans may be necessary so that goals and objectives of restoration and management are articulated and consensus can be reached between stakeholders. The plan integrates ecological, hydrological and human systems both with and in relation to surround lands.

changes to the landscape pattern and processes, the work of ecological engineers through feedback of services and actions controlling self-organization is much needed. The present changing conditions associated with the activities of humans is causing adaptation of the earth's ecosystems to fit them. Ecological engineers can help in this transition by facilitating the movement of species to fill new situations.

Mining, like many other activities of humans, disrupts both organization and driving energies of the landscape which results in ecological restructuring through adaptive self-organization. Throughout the landscape, at the interface between human dominated systems and the natural environment, stand emerging ecosystems. These interface systems often look terrible, disrupted, patchy, stunted and low in productivity. Left alone, nature, using available energy sources and

whatever species are at hand, will eventually develop ecosystems adapted to the new conditions. On the other hand, ecological engineers can assist this adaptive self-organization through their management inputs and facilitate the spread of genetic information that these new emerging environments require.

Since humans have disturbed and displaced large areas of the landscape in which conditions are much different, new species may be required to form new ecosystems. It may be necessary to foster large-scale multiple transplantation of species to develop new designs and new ecosystems. While self-organization and re-establishment of ecological systems will take place anyway, ecological engineers can speed the process by providing new species in tests of adaptive self-organization in carefully selected situations. The mining landscape may need a successional jump-start as

the new conditions associated with mining may require new ecosystems and, therefore, new assemblages of species.

The millions of species of plants, animals and microorganisms are the raw materials of self-design and the palette of the ecological engineer for building new ecosystems. The use of species from different systems generates new combinations of organisms that may self-organize around these changed landscapes. Multiple seeding and the ensuing self-organization, as an ecological engineering technique, may provide new ecosystems for these new conditions, however, sometimes regulation stands in the way of this adaptive self-organization. For instance, the introduction of non-native plants is often forbidden in restoration on the basis that non-natives may be too aggressive and take over large areas to the detriment of native species. In the past, there have such introductions that some feel have resulted in non-natives becoming problematic. The cogon grass (*Imperata cylindrical*) that was first planted in the early 1900's for forage and erosion control is an example.

Some may consider the presence of non-native species in restored landscapes as not true restoration. Yet because the soils of mined lands (the surface parent material) have been altered and in most cases do not resemble those characteristic of native ecological communities, the likelihood of developing ecosystems with the same suite of species found in the native communities is diminished. Add to this, the fact that today there are many more "introduced" species in the Florida landscape than in the past, and it is quite probable that restored phosphate mined lands will contain non-natives. For these reasons, if restored lands are not highly maintained, the presence of non-natives may be impossible to control. Calling non-natives alien species, Ewel and Putz (2004) have suggested that there is a place for them in restoration, and that "blanket condemnation of alien species in restoration efforts is counterproductive". Further they suggest . . .

Risk is always an issue when alien species are involved, but greater risk-taking is warranted where environmental conditions have been severely modified through human activity than where reassembly of biological community is the sole goal of restoration (Ewel and Putz, 2004, p. 354)

Certainly, phosphate-mined lands qualify as severely modified. The new situations created by restoration are dynamic. The flows of energy and organizations of abiotic components often are quite different from conditions prior to the alterations that have lead to restoration. Good restoration under these dynamic conditions recognizes that adaptive self-organization calls for flexibility and that the goal should be to develop functional ecosystems and landscapes which may or may not resemble the ecosystems and landscapes which were found there prior to disruption. In addition, a rigid ideal of what a restored ecosystem or landscape should look like fails to recognize the dynamic nature of restored systems.

4.2. Large-scale restoration as adaptive management

It goes without saying that the phosphate mining district of Florida is a coupled human and natural system. Additionally, it is obvious that there are many actors or stakeholders involved in the landscape transformation of the phosphate district. Scientists, engineers and regulators work directly with shaping restoration efforts, while the citizens who live in and around the district and who are affected by all aspects of the mining system have had significant (and it might be suggested, increasing) impact on present day mining through their interest in and questions concerning the efficacy of restoration. In all, the stakeholders, including scientists, miners, regulators and local citizens are involved in a program of co-evolution of science, management and policy that somewhat parallels the practice of modern resource management known as adaptive management (Holling, 1978; Walters, 1986; Walters and Holling, 1990; Gunderson and Holling, 2002).

Adaptive management is, as the phrase implies, management through adaptively changing interventions as queues from the system that is being managed suggest change is needed. In other words, management that is tightly coupled with and observant of the system that is being managed. A key difference between the co-evolution that has occurred through the 30 years of phosphate mine restoration and what is termed adaptive management is related to how learning occurs. Most of the literature on adaptive management suggests that improving management policies and practices by learning from the outcomes of operational programs

should be based on quantitatively explicit hypotheses about expected system behaviors (Gunderson, 1999; Walters, 1997). In other words, management activities are crafted as *experiments* complete with testable hypotheses to gain knowledge, which implies a formal process of problem assessment, study design, implementation, monitoring, evaluation, and feedback (Nyberg, 1998). Learning that has occurred in the large-scale restoration of phosphate mined lands has occurred by-and-large as a trial and error endeavor in that there is no formal process as outlined above. In fact, because of the diffuse nature of the organization of stakeholders involved, it would be difficult to develop integrated testable hypotheses at the landscape scale.

To be sure, the science behind phosphate restoration research has been driven by testable hypotheses related to the details of restoration. What is lacking in order for the entire regime of landscape restoration after phosphate mining to more closely fit the definition of adaptive management is a formal process where management activities are crafted as experiments. Because of the diffuse nature of restoration research, implementation and land management that is centered in numerous companies, research institutions and governmental agencies, the management process is incremental and often does not formally direct the research process.

It seems to be the case of all large-scale restoration projects that they involve a very diverse and, in some ways, diffuse organization of stakeholders. How to develop successful mechanisms for effective communication and coordination between the various players is a major concern. By the very nature of large-scale restoration programs and the fact that stakeholders are from numerous organizations, it is difficult to act as a coordinated whole and to foster good communication. Yet, adaptive management requires communication and cooperation. Science must understand the overall goal of the management team and develop scientific investigations that reinforce that goal. The management team must understand what the goal is they are striving for, often provided by the regulatory community on behalf of the citizens of the area. Unfortunately, communication over the last 30 years within the combined system of phosphate stakeholders has not been as strong as might be needed for effective restoration.

A key element of adaptive management is learning; some say, learning by doing (Holling, 1978). In 1995, the efforts of Erwin et al. (1997) to evaluate the first 20 years of wetlands restoration in the phosphate districts was an attempt to learn from what had been done in the past. However, learning requires the time to reflect, stop, observe, process and integrate that which has been observed. While a report was written which documented their findings, it appears that it had a little effect. It was quite apparent to the researchers on the Erwin team that wetlands restoration showed marked improvements over the 20 years of active restoration that had thus far been conducted. What was lacking was a broader perspective, one that integrated wetlands and uplands, natural lands and human dominated land uses. They saw that the broader perspective was also missing from hydrologic planning of the watershed. Further, they recognized that a closer working relationship was needed between industry, regulators and citizens. To function as an adaptive management unit, the stakeholders must work together. Adversarial relationships between citizens, industry, scientists and regulators may hinder effective restoration and in fact may be quite detrimental.

Large-scale restoration requires adaptive management. The restoration of phosphate mined lands if treated incrementally, one mine site at a time and one wetland at a time will not result in an integrated landscape, either hydrologically or ecologically. An overarching vision is needed. In addition, overall management and regulation must be adaptive and flexible and must be capable of self-reflection and internal communication. Above all, large-scale restoration may require that all stakeholders be internalized to the process and that they work together as a single management team.

4.3. Phosphate mining restoration compared to Everglades restoration

It has been suggested that Everglades restoration is the largest single restoration program yet undertaken by humanity. In terms of size and dollar amount, no doubt, this is true. However, if one compares the degree of repair necessary, the restoration of phosphate mined lands may be a far greater challenge. In the Everglades, the emphasis is on getting the water right, while in the phosphate districts, it is on

Table 3
Comparison of the restoration of phosphate mined lands with the Everglades restoration

	Phosphate reclamation	Everglades restoration
Total area	0.7 million acres	1.5 million acres
Total cost (in US$)	3.6 billion	7.8 billion
Yearly operating costs (in US$)	0.0	182 million

re-establishing functional landscapes from otherwise barren ones. Table 3 compares Florida phosphate restoration with that of the Everglades. In the end, the area of restored phosphate mined land will equal about half that of the Everglades and its cost almost half as well. Reclaimed phosphate mined lands will be concentrated in the Bone Valley area of central Florida with a lesser area in the north Florida mining district near the Suwannee River. A key difference between the two programs is that the Everglades will cost about US$ 182 million per year to operate, while restored phosphate mined lands should require no operating expenses, although if these lands are managed in the future, there will be some costs associated with land management.

5. Concluding remarks

The large-scale restoration of phosphate mined lands is ecological engineering on a grand scale. It requires an understanding of the adaptive self-organization in ecological systems, the dynamics of political systems and the demands of social systems. The following are several essentials we have learned from the evolving, adaptively managed system of phosphate mining and restoration:

- Restoration research has stressed the importance of developing complete ecosystems that incorporate all strata, over-story, mid-story and under-story vegetation ... with the understanding that the resulting ecosystems would then develop a full compliment of fauna. But it appears that successful ecological restoration is often only judged by survival of a single strata. There have been few, if any, attempts, to develop complete ecosystems on mined lands.

- Research results have shown the beneficial role of all early successional, pioneer species as necessary parts of ecological self-organization, yet there remains a relatively stubborn belief that pioneer species are undesirable and must be eliminated from reclaimed ecosystems.
- The requirement for type-for-type ecosystem restoration may be counter to what is needed in large-scale restoration. The conditions in newly constructed landscapes often do not resemble conditions prior to disruption and, therefore, expecting that replacement of type-for-type may be unrealistic.
- The branding of some species as undesirable because they are non-native to the region of restoration may be limiting restoration potential and restoration success. New conditions call for new assemblages of species.
- Large-scale restoration requires large-scale planning. Without an overall vision and goal, the resulting landscape will suffer from disjointed incrementalism and the political and social systems that revolve around the landscape will continue to be at odds with direction and potential outcomes they cannot perceive.

We have learned much regarding restoration of drastically altered lands. Most of what has been learned is related to the details ...what species are characteristic of native ecosystems, the benefits of mulching related to ecosystem development, wildlife use of restored lands, hydrology of mined lands and so forth. Future research should address the larger scale issues related to adaptively managing the restoration landscape and how it will be hydrologically and ecologically organized and how humans will fit within that organization. Future management need be visionary. Since restoration is incremental and phased as lands are mined, it needs to have a vision of the whole that can be articulated and understood by stakeholders and that can then drive both adaptively managed restoration and the science that is driving it.

Appendix A. Definitions

Ecological restoration: The process of assisting re-establishment of a natural community, habitat, species population, or other ecological attribute that

has been eliminated or greatly reduced on a given location.

Ecosystem functions: The dynamic attributes of ecosystems, including interactions among organisms and interactions between organisms and their environment (http://www.ser.org/content/ecological_restoration_primer.asp#3).

Landscape: A heterogeneous land area composed of two or more ecosystems that exchange organisms, energy, water and nutrients.

Landscape restoration: Assisting the re-establishment of ecological and hydrological functions of landscapes composed of two or more ecosystems.

Hydrologic function: The dynamic attributes of the hydrologic systems of an area including base flow, overland flow, recharge, energy transfer and flooding regime and how these attributes affect nutrient cycling, water quality, and aquatic and terrestrial life.

Disjointed incrementalism: A policy making process which produces decisions only marginally different from past practice (after Lindblim, 1959).

Beneficiation: The process of separating a wanted mineral from other material that also is contained in the matrix. In the case of phosphate, this means separating clay and sand from the phosphate rock. A mechanical process called washing is used to separate the larger phosphate pebbles from the ore. A process called flotation is used to recover the finer particles of phosphate from sand (www.imcglobal.com/general/education_corner/phosphates/terms.htm).

Reclamation: The process by which lands disturbed as a result of mining activity are reclaimed back to a beneficial land use. Reclamation activity includes the removal of buildings, equipment, machinery, other physical remnants of mining, closure of settling areas and impoundments and other mine features, and contouring, covering and revegetation of disturbed areas.

Overburden: Layers of soil or rock overlaying a deposit of useful materials or ores. In surface mining operations, overburden is removed using large equipment piled in spoil piles and later used to backfill areas previously mined or in the construction of clay settling area dikes.

Phosphate matrix: A mixture of phosphate pebbles, sand and clay found about 10 m beneath the ground surface.

Appendix B. Rules pertaining to landscape restoration as set forth by the Bureau of Mine Reclamation in Chapter 62C-16, Florida Administrative Code

B.1. 62C-16.0051 Reclamation and Restoration Standards

This section sets forth the minimum criteria and standards which must be addressed in an application for a program to be approved.

(1) *Backfilling and contouring*: The proposed land use after reclamation and the types of landforms shall be those best suited to enhance the recovery of the land into mature sites with high potential for the use desired.
   (a) Slopes of any reclaimed land area shall be no steeper than 4 ft horizontal to one foot vertical to enhance slope stabilization and provide for the safety of the general public.
(2) *Soil zone*
   (a) The use of good quality topsoils is encouraged, especially in areas of reclamation by natural succession.
   (b) Where topsoil is not used, the operator shall use a suitable growing medium for the type vegetative communities planned.
(3) Wetlands which are within the conceptual plan area which are disturbed by mining operations shall be restored at least acre-for-acre and type-for-type.
(4) *Wetlands and water bodies*: The design of artificially created wetlands and water bodies shall be consistent with health and safety practices, maximize beneficial contributions within local drainage patterns, provide aquatic and wetlands wildlife habitat values, and maintain downstream water quality by preventing erosion and providing nutrient uptake. Water bodies should incorporate a variety of emergent habitats, a balance of deep and shallow water, fluctuating water levels, high ratios of shoreline length to surface area and a variety of shoreline slopes.
   (a) At least 25% of the highwater surface area of each water body shall consist of an annual zone of water fluctuation to encourage emergent and transition zone vegetation. This area will also qualify as wetlands under the requirements of

subsection (4) above, if requirements in paragraph 62C-16.0051(9)(d) are met. In the event that sufficient shoreline configurations, slopes or water level fluctuations cannot be designed to accommodate this requirement, this deficiency shall be met by constructing additional wetlands adjacent to and hydrologically connected to the water body.

(b) At least 20% of the low water surface shall consist of a zone between the annual low water line and 6 ft below the annual low water line to provide fish bedding areas and submerged vegetation zones.

(c) The operator shall provide either of the following water body perimeter treatments of the high water line:

1. A perimeter greenbelt of vegetation consisting of tree and shrub species indigenous to the area in addition to ground cover. The greenbelt shall be at least 120 ft wide and shall have a slope no steeper than 30 ft horizontal to one foot vertical.

2. A berm of earth around each water body which is of sufficient size to retain at least the first one inch of runoff. The berm shall be set back from the edge of the water body so that it does not interfere with the other requirements of subsection (5).

(5) *Water quality*

(a) All waters of the state on or leaving the property under control of the taxpayer shall meet applicable water quality standards of the Florida Department of Environmental Protection.

(b) Water within all the wetlands and water bodies shall be of sufficient quality to allow recreation or support fish and other wildlife.

(6) *Flooding and drainage*

(a) The operator shall take all reasonable steps necessary to eliminate the risk that there will be flooding on lands not controlled by the operator caused by silting or damming of stream channels, channelization, slumping or debris slides, uncontrolled erosion or intentional spoiling or diking or other similar actions within the control of the operator.

(b) The operator shall restore the original drainage pattern of the area to the greatest extent possible. Watershed boundaries shall not be crossed in restoring drainage patterns; watersheds shall be restored within their original boundaries. Temporary roads shall be returned at least to grade where their existence interferes with drainage patterns.

(7) *Revegetation*: The operator shall develop a revegetation plan to achieve permanent revegetation which will minimize soil erosion, conceal the effects of surface mining, and recognize the requirements for appropriate habitat for fish and wildlife.

(a) The operator shall develop a plan for the proposed revegetation, including the species of grasses, shrubs, trees, aquatic and wetlands vegetation to be planted, the spacing of vegetation and, where necessary, the program for treating the soils to prepare them for revegetation.

(b) All upland areas must have established ground cover for 1 year after planting over 80% of the reclaimed upland area, excluding roads, groves or row crops. Bare areas shall not exceed one-quarter (1/4) acre.

(c) Upland forested areas shall be established to resemble pre-mining conditions where practical and where consistent with proposed land uses. At a minimum, 10% of the upland area will be revegetated as upland forested areas with a variety of indigenous hardwoods and conifers. Upland forested areas shall be protected from grazing, mowing or other adverse land uses to allow establishment. An area will be considered to be reforested if a stand density of 200 trees/acre is achieved at the end of 1 year after planting.

(d) All wetland areas shall be restored and revegetated in accordance with the best available technology.

1. Herbaceous wetlands shall achieve a ground cover of at least 50% at the end of 1 year after planting and shall be protected from grazing, mowing or other adverse land uses for 3 years after planting to allow establishment.

2. Wooded wetlands shall achieve a stand density of 200 trees/acre at the end of 1 year after planting and shall be protected from grazing, mowing or other adverse land uses

for 5 years or until such time as the trees are 10 ft tall.

(e) All species used in revegetation shall be indigenous species except for agricultural crops, grasses and temporary ground cover vegetation.

References

Brown, M.T., Bardi, E., 2001. Emergy of ecosystems. In: Emergy. Folio 4. of Handbook of Emergy Evaluation. The Center for Environmental Policy, University of Florida, Gainesville, 93 pp.

Brown, M.T., Carstenn, S., Reiss, K., 2000. Evaluation of Successional Trajectories in Constructed Wetlands on Phosphate Mined Lands. Report to the Florida Institute of Phosphate Research. Center for Wetlands, University of Florida, Gainesville, FL.

Brown, M.T., Carstenn, S., Jackson, K., Bukata, B.J., Sloane, M., 1998. Shading Effects on Nuisance Species. Final Report to Florida Institute of Phosphate Research. Center for Wetlands, University of Florida, Gainesville, FL.

Brown, M.T., Odum, H.T., Gross, F., Higman, J., Miller, M., Diamond, C., 1985. Studies of a Method of Wetland Reconstruction following Phosphate Mining. Final Report to the Florida Institute of Phosphate Research, Bartow, Florida. Center for Wetlands, University of Florida, Gainesville, Florida. 76 pp.

Brown, M.T., Tighe, R.E. (Eds.), 1991. Techniques and Guidelines For Reclamation of Phosphate Mined Lands. Florida Institute of Phosphate research, Bartow, FL, p. 386, Publication no. 03-044-095.

Cates, J.W.H., 1992. A Regional Conceptual Plan for The Southern Phosphate District of Florida. Bureau of Mine Reclamation, Florida Department of Environmental Protection, Tallahassee, FL, 179 pp.

Ewel, J., Putz, F., 2004. A place for alien species in ecosystem restoration. Front Ecol. Environ. 297, 354–360.

Erwin, K.L., Doherty, S.J., Brown, M.T., Best, G.R. (Eds.), 1997. Evaluation of Constructed Wetlands on Phosphate Mined Lands in Florida. Florida Institute of Phosphate research, Bartow, FL, p. 698.

Florida Institute of Phosphate Research, 2004, http://www.fipr.state.fl.us/southintro.htm.

Florida Phosphate Council, 2004. Florida Phosphate Facts, Florida Phosphate Council, 1435 East Piedmont Drive, Suite 211 Tallahassee, FL, 8 pp.

Holling, C.S. (Ed.), 1978. Adaptive Environmental Assessment and Management. John Wiley & Sons, New York.

Gunderson, L., 1999. Resilience, flexibility and adaptive management—antidotes for spurious certitude? Conserv. Ecol. 3 (1), 7, URL: http://www.consecol.org/vol3/iss1/art7/.

Gunderson, L.H., Holling, C.S., 2002. Panarchy: Understanding Transformations in Human and Natural Systems. Island Press, Washington, 507 pp.

Gunderson, L.H., Holling, C.S., Light, S., 1995. Barriers and Bridges to Renewal of Ecosystems and Institutions. Columbia University Press, New York, NY.

Laszlo, E., 1972. Introduction to Systems Philosophy. Harper and Row, New York, 328 pp.

Lindblim, C.E., 1959. The Science of "Muddling Through". Public Adm. Rev. 19 (Spring), 79–88.

Long Jr., H.W., Orne, D.P., 1990. Regional Study of Land Use Planning and Reclamation. Florida Institute of Phosphate Research, Bartow, FL.

Nyberg, J.B., 1998. Statistics and the practice of adaptive management. In: Sit, V., Taylor, B., (Eds.), Statistical Methods for Adaptive Management Studies, Land Manage. Handbook 42, B.C. Ministry of Forests, Victoria BC, pp. 1–7.

Odum, H.T., 1996. Environmental Accounting: Emergy and Environmental Decision Making. John Wiley and Sons, New York, 370 pp.

Odum, H.T., Miller, M.A., Rushton, B.T., McClanahan, T.R., Best, G.R., 1983. Interactions of Wetlands with the Phosphate Industry. Final report. Florida Institute of Phosphate Research, Publication No. 03-007-025.

Richardson, S.G., Kluson, R.A., 2000. Managing nuisance plant species in forested wetlands on reclaimed phosphate mined lands in Florida. In: Cannizzaro, P.J. (Ed.), Proceedings of the 26th Annual Conference on Ecosystems Restoration and Creation. Hillsborough Community College, Tampa, FL, May, 1999, pp. 104–118.

Robertson, D.J., 1985. Freshwater Wetland Reclamation in Florida: An Overview. The Florida Institute of Phosphate Research, 1855 West main Street, Bartow, FL.

Rushton, B.T., 1988. "Wetland reclamation by accelerating succession." Ph.D. Dissertation, Gainesville, FL, University of FL, 266 pp. (CFW-88-01).

Walters, C., 1986. Adaptive Management of Renewable Resources. Macmillan, New York.

Walters, C., 1997. Challenges in adaptive management of riparian and coastal ecosystems. Conserv. Ecol. 1 (2), 1,, URL: http://www.consecol.org/vol1/iss2/art1/.

Walters, C.J., Holling, C.S., 1990. Large-scale management experiments and learning by doing. Ecology 71, 2060–2068.

Weber, T.C., 1994. Spatial and temporal simulation of forest succession with implications for management of bioreserves. M.S. Thesis. Department of Environmental Engineering Sciences. University of Florida, Gainesville, 258 pp.

Available online at www.sciencedirect.com

SCIENCE  DIRECT°

ELSEVIER   Ecological Engineering 24 (2005) 331–340

ECOLOGICAL ENGINEERING

www.elsevier.com/locate/ecoleng

Are we purveyors of wetland homogeneity?
A model of degradation and restoration to improve wetland mitigation performance

Robert P. Brooks[*], Denice Heller Wardrop, Charles Andrew Cole[1], Deborah A. Campbell[2]

Penn State Cooperative Wetlands Center, Department of Geography, 302 Walker Building, Pennsylvania State University, University Park, PA 16802, USA

Received 29 April 2004; received in revised form 14 July 2004; accepted 22 July 2004

Abstract

The national goal of no net loss of wetland functions is not being met due to a variety of suboptimal policy and operational decisions. Based on data used to develop a conceptual model of wetland degradation and restoration, we address what can be done operationally to improve the prospects for replacing both the area and functions of mitigated wetlands. We use measures of hydrologic, soil, and biodiversity characteristics from reference standard sites, degraded wetlands, and created wetlands to support our premise. These data suggest that wetland diversity and variability often become more homogeneous when subjected to a set of stressors. The degradation process reduces the original heterogeneity of natural wetlands. In addition, soil characteristics and composition of biological communities of creation projects may mirror those of degraded wetlands. We recommend that scientists and managers use identical sampling protocols to collect data from reference wetlands that can be used to assess the condition of degraded wetlands and to improve the design and performance of mitigation projects.
© 2005 Elsevier B.V. All rights reserved.

Keywords: Wetlands; Restored wetlands; Created wetlands; Mitigation; Reference sites

1. Introduction

1.1. Wetland mitigation policy and practice

The U.S. Environmental Protection Agency's Office of Wetlands, Oceans, and Watersheds established two national priorities for wetlands in 2000: for states and tribes to develop wetland monitoring programs, and to improve the success rate of compensatory mitigation

* Corresponding author. Tel.: +1 814 863 1596;
fax: +1 814 863 7943.
    E-mail address: rpb2@psu.edu (R.P. Brooks).
[1] Present address: Center for Watershed Stewardship, Hamer/Heinz Building, Pennsylvania State University, University Park, PA 16802, USA
[2] Present address: EDAW Inc., 240 East Mountain Avenue, Fort Collins, CO 80524, USA

0925-8574/$ – see front matter © 2005 Elsevier B.V. All rights reserved.
doi:10.1016/j.ecoleng.2004.07.009

(D. Vetter, personal communication). In 2002, federal resource agencies released a National Wetlands Mitigation Action Plan (U.S. Army Corps of Engineers, 2002) designed to improve the ecological performance of compensatory wetland mitigation. The Action Plan acknowledged the critical evaluations of wetland mitigation performance by the National Research Council (NRC, 2001) and the U.S. General Accounting Office (2002), and affirmed a commitment to a goal of no net loss of wetlands through improved accountability, monitoring, and research. Clearly, there is a renewed and strong commitment to assessing and restoring wetlands as regulated "waters of the U.S.", and to do so on a watershed basis (e.g., http://www.epa.gov/owow/). The prospects for comprehensive wetlands monitoring, assessment, and mitigation have never been stronger, either from regulatory or technical perspectives. As stated by the National Research Council (NRC, 2001), the goal of no net loss of wetland functions is not being met due to a variety of suboptimal policy and operational decisions. In this paper, by way of a conceptual model of wetland degradation and restoration, operational decisions that can greatly improve the prospects for replacing both the area and functions of wetlands are addressed.

There are two primary reasons to engage in wetland restoration and creation projects. To comply with environmental regulations, there is an expectation of compensatory mitigation for losses of wetland area or function. Losses to wetlands that occur as a result of a permitting process usually are required to be mitigated. Typically, this mitigation results in a project designed to replace at least the same amount of wetland area impacted. Replacement of specified functions originally provided by that impacted wetland also might be required. Implementing the project becomes a condition of the issued permit.

Outside the regulatory arena, an interested group or person elects to design and implement a project that is perceived to increase or enhance wetland resources (NRC, 1992). These types of projects are usually initiated voluntarily. In either case, each project proceeds through a series of steps that includes selection of a suitable site, acquisition or gaining access to the chosen site, development of conceptual designs, preparation of construction specifications and implementation plans, and eventual construction of the wetland (Brooks, 1993). Post-construction monitoring may or may not be required or occur.

The degree to which wetland restoration and creation projects achieve some measure of success has received considerable debate (Kusler and Kentula, 1990; Brooks, 1993; Brinson and Rheinhardt, 1996; Mitsch and Wilson, 1996; Cole et al., 1997a; Zedler and Callaway, 1999). Proponents tout the potential benefits, real or perceived, of increasing wetland area and function toward an overall net gain in the resource. Opponents argue that mitigation is a license to impact natural wetlands, and that the resultant projects have scant resemblance to the wetlands they are supposed to replace.

1.2. A model of wetland degradation and restoration

Based on comparative studies of reference and created wetlands (e.g., Bishel-Machung et al., 1994; Cole and Brooks, 2000; Campbell et al., 2002) and our collective experience of comparing reference wetlands from a range of hydrogeomorphic (HGM) subclasses along a disturbance gradient (e.g., Cole et al., 1997b; Wardrop and Brooks, 1998; O'Connell et al., 2000; Brooks, 2004; Brooks et al., 2004), we are proposing a conceptual model of wetland degradation and restoration as a set of testable hypotheses. We challenge scientists and managers to examine our model, test the associated hypotheses, and if valid, make the necessary adjustments to reduce further degradation of natural wetlands and to improve the performance of wetland restoration and creation projects.

The conceptual model, presented schematically in Fig. 1, recognizes a degradation process caused by one or more stressors. The typical range of stressors affecting wetlands was summarized by Adamus and Brandt (1990) and expanded by Adamus et al. (2001). The cumulative effects of any combination of these stressors appear to result in a convergence of characteristics producing a more homogeneous group of wetlands. In turn, those degraded wetlands are most similar to created sites.

1.3. Hypotheses

Our four related hypotheses are:

Conceptual Model of Wetland Degradation and Restoration

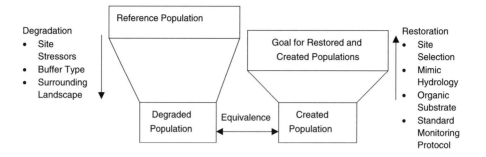

Fig. 1. Model of wetland degradation and restoration showing the equivalence of degraded and created population characteristics, and the goal of mimicking reference wetlands with mitigation projects.

1. Wetlands change in structure and function when subjected to one or more stressors resulting in a recognized degradation sequence

   The ultimate form of degradation is complete loss of areal extent (e.g., fill or dewatering) at which time the wetland ceases to exist in that location. Inventory trend data have been reported primarily with regard to changes in areal extent (e.g., Dahl, 1990). Much more widespread, yet more difficult to quantify, has been the degradation of wetlands, defined as loss of functional performance. The application of HGM assessment models or similar approaches provides a means to quantify loss of wetland function (Brinson, 1993; Rheinhardt et al., 1999).

2. Degraded wetlands develop characteristics that differ from reference standard sites that can be expressed as a change in condition

   One option in wetland mitigation is to create replacement wetlands. Questions have been raised about whether created wetlands are equivalent in structure and function to the natural wetlands they replace (Zedler and Callaway, 1999). Galatowitsch and van der Valk (1994) stated that the definitive test of mitigation success was how well restored wetlands resemble natural wetlands; the same could be said of created sties. Our findings (Bishel-Machung et al., 1994; Cole and Brooks, 2000; Campbell et al., 2002) agree with general statements made by NRC (2001) that created wetlands may never express the full range of ecological variability found in natural wetlands. Many created and restored sites are far wetter than natural wetlands, with extensive areas of open water.

3. Substantial improvements in structural and functional performance of created and restored wetlands can occur if data from reference wetlands are used to design mitigation projects and to evaluate their success

   It is imperative that criteria derived from studies of reference wetlands be used in the design, construction, and evaluation of mitigation projects. In addition, to optimize the use of data through the assessment, design, and evaluation processes, consistent use of a consistent sampling protocols when comparing among natural wetlands and projects is essential (e.g., Gray et al., 1999; Wardrop et al., 2004).

4. A wetland mitigation project can match the structure and function of a reference wetland in a comparable HGM subclass for a given ecoregion

   We do not believe that exact replication of natural wetland ecosystems is possible, but we should aspire to approach the goal of replicating structure and function using the best possible designs and construction methods based on the best possible science derived from studies of reference wetlands and other experimental work. This assumes, of course, that current or historical data from reference sites can be acquired.

2. Methods

The use of reference sites has become increasingly more common as ecologists and managers search for reasonable and scientifically based methods to measure

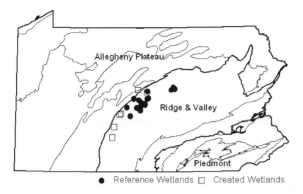

Fig. 2. Locations of reference wetlands and created wetlands compared in Pennsylvania.

and describe the inherent variability in natural aquatic systems (Hughes et al., 1986; Kentula et al., 1992; Brinson, 1993; Rheinhardt et al., 1999). Although reference sites often represent areas of minimal human disturbance (i.e., reference standards in HGM wetlands parlance; Smith et al., 1995), in many instances it is more useful to represent a range of environmental conditions across a landscape. Classification of reference wetlands in the HGM approach harnesses wetland variability, and when integrated with the wetland classification scheme, provides a framework to characterize observed differences in wetland structure and function.

From 1993–2003, the Penn State Cooperative Wetlands Center (CWC) compiled a total of 222 reference wetlands across the five major ecoregions of Pennsylvania and spanning seven HGM subclasses. HGM subclasses were based on a regional classification system for Pennsylvania and adjoining ecoregions (Cole et al., 1997a,b; Brooks, 2004). The original intent for the reference set was to use reference data to improve the design and construction of restoration and creation projects, but aspects of the HGM approach were rapidly assimilated into the investigations of our reference wetlands to facilitate functional comparisons and condition assessments (Brooks, 2004; Brooks et al., 2004). We gleaned representative data used in this paper from this reference set and coupled it with comparable data from created wetlands (Bishel-Machung et al., 1996; Campbell, 1996; Cole and Brooks, 2000; Campbell et al., 2002) (Fig. 2). Riparian depressions are groundwater-supported sites with closed contours and a single outlet, usually located at the base of a hillslope near streams. Slopes have unidirectional flow

along a topographic gradient, and are generally supported by a mixture of surface runoff and groundwater. Headwater floodplains are influenced by the overbank flooding regime of second order streams and by overland flow (Cole et al., 1997a,b).

Categorizing wetlands as degraded was based on a human disturbance score that combines landscape, buffer, and on-site stressors (Brooks et al., 2004). Observed stressors are compiled into general categories as described by Adamus and Brandt (1990) and Adamus et al. (2001). Wetlands in Pennsylvania are most often affected by hydrologic modifications, sedimentation, and alteration of natural vegetation.

We present the mean and S.D. for most variables, computed on a site basis, to indicate how degraded wetlands and created sites differ from reference standard wetlands. For most categories of HGM subclass and wetland type, the number of sites is low, so no statistical comparisons were attempted. Rather, our intent is to suggest trends worthy of further investigation, and to encourage use of reference site data to improve mitigation design, construction, and performance.

3. Results

Based on our collective work and a review of other selected papers, we believe hypotheses 1 and 2 can be accepted. In addition to offering supportive citations, we present selected data that supports both the degradation and the homogeneity aspects of the conceptual model.

3.1. Condition of reference wetlands along a disturbance gradient

Although land use patterns surrounding wetlands do not completely describe the level of observed disturbance or degradation found in wetlands, they are usually highly correlated (e.g., Wardrop and Brooks, 1998; O'Connell et al., 2000; Brooks et al., 2004). When wetlands are characterized during field studies, we have identified significant changes in wetland structure and function tied to stressors emanating from human-induced disturbances in the surrounding landscape. For the purposes of this paper, we have selected a few illustrative measures collected from 23 natural wetlands in three HGM subclasses; riparian depressions,

Table 1
Reference standard, degraded, and created wetlands studied in Pennsylvania

Site number	CWC number	CWC site name	County	HGM subclass	Years sampled (created)
1	5*	McCall Dam	Centre	Riparian depression	1993, 1994
2	6*	Sand Spring	Union	Riparian depression	1993, 1994
3	10*	Whipple Dam SP	Centre	Riparian depression	1993, 1994
4	13*	Clark's Trail	Union	Riparian depression	1993, 1994
5	18B	Buffalo Run	Centre	Riparian depression	1993, 1994
6	52	Tadpole	Centre	Riparian depression	1997
7	56	Farm 12	Centre	Riparian depression	1997
8	59	NBB–RD	Centre	Riparian depression	1997
9	1	BESP-PFO	Centre	Slope	1993, 1994
10	2	BESP-PFO	Centre	Slope	1993, 1994
11	14	LFC–PFO	Centre	Slope	1993, 1994
12	19*	Rothrock State Forest	Huntingdon	Slope	1993, 1994
13	24*	McGuire Rd	Huntingdon	Slope	1994
14	25	Windy Hill Farms	Centre	Slope	1994
15	54	Wardrop's	Centre	Slope	1994
16	55	Swamp White Oak	Centre	Slope	1997
17	4*	LFC Dam	Centre	Headwater floodplain	1993, 1994
18	18A	Buffalo Run	Centre	Headwater floodplain	1993, 1994
19	26	Water Authority	Centre	Headwater floodplain	1994
20	31	Cedar Run	Centre	Headwater floodplain	1994
21	53	NBB–HWF	Centre	Headwater floodplain	1997
22	57	Thompson Run	Centre	Headwater floodplain	1997
23	60	Laurel Run	Huntingdon	Headwater floodplain	1997
24	C1	Rt. 220A	Blair	Created	1995 (1993)
25	C2	Peterson Industrial Park A	Blair	Created	1995 (1992)
26	C5	Tipton	Blair	Created	1995 (1991)
27	C6	Snowshoe	Centre	Created	1995 (1990)
28	C7	Mt. Eagle	Centre	Created	1995 (Late 1980s)
29	C11	Sproul Interchange	Blair	Created	1995 (Late 1970s)
30	C12	Duncansville	Blair	Created	1995 (Late 1970s)

(*) Any site that is a reference standard.

slopes, and headwater floodplains (Cole et al., 1997b; Brooks, 2004) (Table 1). We provide examples from hydrologic, biogeochemical, and biodiversity functional categories, which typically are addressed in HGM assessment models (Smith et al., 1995) and indices of biological integrity (Karr and Chu, 1999).

We monitored water levels in reference wetlands and found that hydropatterns differed by HGM subclasses and changed with disturbance (Cole et al., 1997b) (Table 2). Sedimentation rates were significantly greater in disturbed sites compared with reference standards (Wardrop and Brooks, 1998). Morphological characteristics of soils differed as well; organic matter declined with disturbance and soil chroma suggested drier conditions prevailed in degraded sites (Campbell et al., 2002) (Table 2).

The CWC has conducted a number of studies addressing the response of various biological taxa to human induced disturbances. Field studies showed reduced richness in vascular plants and increases in the dominance of invasive plants (Campbell et al., 2002) (Table 2, Fig. 3). Greenhouse experiments designed to mimic stressors observed in the field suggest mechanisms that might be responsible for these trends. Walls et al. (2004) showed that sedimentation rates above expected amounts inhibit the germination and growth of riparian trees seedlings. Mahaney et al. (2004) found that sediment and nutrient stresses on herbaceous hydrophytes could alter plant community composition in favor of aggressive invasive species such as reed canarygrass (*Phalaris arundinacea*), which was comparable to findings by Green and Galatowitsch (2002) and Kercher and Zedler (2004). Wetland macroinvertebrate taxa were sampled in selected wetlands, and the data were compiled into a macroinvertebrate index of community integrity (ICI) (Bennett

Table 2
Selected data comparisons among reference, degraded, and created wetlands

Variable	Wetland type						
	Riparian depression, $n = 8$		Slope, $n = 8$		Headwater floodplain, $n = 7$		Created $n = 7$
	Reference	Degraded	Reference	Degraded	Reference	Degraded	
Hydrology							
Median depth (cm)	−10	+9	−23	−18	−15	−42	−8
Percent time root zone	81 ± 19	83 ± 20	66 ± 36	59 ± 19	91 ± 12	35 ± 20	78 ± 31
Percent time inundated	13 ± 11	77 ± 27	16 ± 17	3 ± 1	1 ± 1	5 ± 4	35 ± 26
Percent time dry	19 ± 19	18 ± 20	36 ± 35	41 ± 19	90 ± 56	58 ± 21	22 ± 31
Percent time saturated	68 ± 25	7 ± 8	49 ± 31	56 ± 20	53 ± 57	37 ± 18	42 ± 23
Soils							
Chroma (Munsell color)	1.1 ± 0.2	2.0 ± 0.1	1.5 ± 0.6	1.9 ± 0.5	1.2 ± 0.1	2.1 ± 0.5	2.3 ± 0.5
Percent organic matter	24 ± 9	9 ± 8	21 ± 13	8 ± 1	13 ± 2	6 ± 2	4 ± 1
Biodiversity							
Proportion of exotic and invasive plants	0 ± 0	0.35 ± 0.19	0.15 ± 0.19	0.40 ± 0.23	0.10 ± 0.14	0.48 ± 0.23	0.57 ± 0.24
Macroinvertebrate index of community integrity	31 ± 6	9 ± 2	33 ± 6	18 ± 3	NA	NA	20 ± 5
Bird community index	51 ± 23	39	57 ± 7	40 ± 6	NA	35 ± 5	NA

Median depth (cm)—median depth above (+) or below (−) ground level for all measurements taken in slotted wells; Percent time root zone—percent of total observations from slotted wells where water was within 30 cm of the ground surface; percent time inundated—percent of total observations from slotted wells where water was above the ground surface; percent time dry—percent of total observations from slotted wells where no water was recorded in slotted wells, regardless of absolute depth; percent time saturated—percent of total observations from slotted wells where water was recorded in slotted wells within 10 cm of the ground surface; chroma (Munsell color)—color recorded from Munsell Color chart, calculated as a mean value per wetland; percent organic matter—determined as loss on ignition of oven-dried samples at 450 °C (Storer, 1984); proportion of exotic and invasive plants—proportion of exotic and invasive vascular plant species in the most dominant species per wetland (Brooks, 2004); macroinvertebrate index of community integrity—based on multi-metric index (score range of 5–45) developed by Bennett (1999); bird community index—based on multi-metric index (score range of 20–77) developed by O'Connell et al., (2000); NA—indicates that data were not available.

and Brooks, unpublished). ICI scores were lower for disturbed wetlands and created sites. Similarly, songbirds were sampled across the wetland disturbance gradient, encompassing the surrounding landscape, and assembled into a bird community index (BCI) (O'Connell et al., 2000). BCI scores were based on bird guilds that were generated independently from wetland or landscape characteristics. Degraded wetlands had lower BCI scores than reference standard sites (Table 2). Created wetlands were not sampled for birds.

3.2. Reference wetlands versus created wetlands

During previous CWC research projects, selected reference wetlands were compared to several populations of wetland mitigation sites (Bishel-Machung et al., 1996; Campbell, 1996; Cole and Brooks, 2000; Campbell et al., 2002). We selected seven created wet-

lands where sufficient data were available to compare to the aforementioned 23 natural wetlands (Table 1).

Mitigation sites had higher amounts of sand and gravel, and lower amounts of organic matter, silt, and clay. Wetland mitigation sites contained more large particles in soils than reference wetlands. The presence of large amounts of large particles reflects construction practices that may involve excavation by blasting, or removal of upper layers of soil. Soil bulk density was higher in mitigation sites, again, a reflection of construction practices (compaction by machinery). Bulk density was inversely correlated with organic matter content. Soil matrix chroma, used as an indicator of the extent of soil saturation, was also higher than in reference wetlands, indicating that the imposed hydrologic regimes of created wetlands were insufficient to generate saturated or inundated conditions, which would facilitate iron reduction leading to greyer colors (Table 2). High chromas also reflect low levels of organic matter,

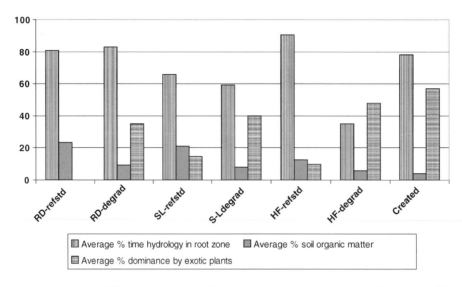

Fig. 3. Comparisons among wetland variables (percent time water in root zone, percent organic matter in soil, proportional dominance of exotic and invasive plant species) for reference standard (RefStd), degraded (Degrad) and created (Created) wetlands from three hydrogeomorphic subclasses; riparian depression (RD), slope (SL), and headwater floodplain (HF).

which normally functions as a substrate for reducing bacteria.

Campbell et al. (2002) found that created wetlands were more similar to degraded natural wetlands, than reference standard wetlands. These created sites ranged in age from 1–18 years since construction, thus, questions about lag times or evolution of restoration and creation technology were partially answered. The small amounts of organic matter found in mitigation sites could simply be due to project age, since accretion rates are slow and none of the mitigation sites were older than approximately 18 years at the time of the study. Created wetlands >10 years of age has significantly more organic matter than sites <10 years of age, although both had significantly less organic matter than reference standard or degraded wetlands (Campbell et al., 2002). Whether this age-based difference was due either to differences in time of construction or construction techniques is not known. It does not appear, however, that organic matter will accumulate within a decade to levels typical of natural wetlands. Average amounts of organic matter were >10% for the reference standard sites in the three subclasses of natural wetlands studied, whereas both degraded and created wetlands averaged <10% (Table 2, Fig. 3).

Vascular plant richness and total cover were both greater in reference versus created wetlands (Campbell et al., 2002). The proportion of dominant plants that were invasive was substantially greater in created wetlands, when compared with both reference standard and disturbed natural wetlands (Table 2, Fig. 3). Reference wetlands had a more complex perimeter to area relationship than in mitigation sites, indicating that there is a tendency to create regular, geometric shapes during the wetland construction because they are less expensive and are simpler to build (Campbell, 1996).

4. Discussion

In this paper, we provide evidence that regardless of original intent, wetland mitigation projects result in wetlands of moderate to low condition, that are in some ways, structurally and functionally equivalent to moderately and severely degraded natural wetlands. When compared to natural reference standard wetlands (e.g., Smith et al., 1995), or those approaching the best possible condition in a given region, creation projects emulate degraded wetlands in their soil characteristics and dominance by invasive plants. With regard to hydrology, created sites span a continuum from being too dry

to be considered a jurisdictional wetland, to being inundated with water that is too deep to support emergent and woody plant communities typical of natural wetlands. Created wetlands that are excessively inundated become divergent from severely degraded wetlands as well. The end result in both cases is often a more homogeneous set of wetlands that does not resemble their original natural counterparts. The range of natural variability is missing in these two sets of wetlands; mitigation projects and degraded wetlands (Fig. 1).

As suggested by Sibbing (2003), fundamental changes in wetland mitigation policies and practices are needed to remedy this trend. If managers and practitioners truly intend to restore lost area and function that encompasses the full diversity of wetland types, then steps must be taken to improve the design and construction models that are currently in use. We propose that all restoration and creation projects be based on the structural and functional characteristics of natural reference standard wetlands of the same HGM and vegetation type for any given geographic region. Within this context, we refer to these specific characteristics as performance criteria, because ultimately, the same measures used to assess wetland conditions should be used in the design and construction process, and during evaluation of success. That is, at the conclusion of the project, does the wetland perform as intended? By using the same protocols throughout permitting, design, construction, and monitoring processes, restoration and creation projects are more likely to succeed in mimicking or at least replacing equivalent structures and functions found in wetlands of high ecological integrity for any given wetland type.

Based on these collective results, we developed the illustrative model (Fig. 1) that suggests how wetland diversity and variability often become more homogeneous when subjected to a set of stressors. There is a convergence of characteristics such that the end result is a population of wetlands that are relatively homogeneous. The original variability expressed as a set of heterogeneous measures from less disturbed natural wetlands has been lost. In addition, the soil characteristics and composition of biological communities of creation projects may mirror those of degraded wetlands. Hydrologically, created wetlands show wide variability in their hydrologic regimes, which can cause them to be either similar to degraded wetlands, or appear in a unique subclass of their own.

The implications are clear. The inherent heterogeneity of naturally occurring wetlands is lost as degradation progresses. Functions attributed to a diverse set of wetland types are necessarily reduced, resulting in lower species richness and changes in species composition across multiple taxa (e.g., vascular plants, macroinvertebrates, and birds). When these degraded wetlands are compared to created projects, they may be equivalent or form their own unique cluster. This indicates that there is room for considerable improvement in the design and construction of created wetlands, and probably restored wetlands, too. If these projects are structurally equivalent to degraded wetlands, then they undoubtedly provide comparable levels of function, and possibly fewer numbers of functions.

The conceptual model we propose suggests three, intertwined courses of action for managers and practitioners. All three rely heavily on the use of reference standard wetlands as design templates for protection, restoration, and creation projects. (1) We should focus our attention not only on losses of wetland area, but also losses of functions. This suggests a need for greater protection of existing wetlands of relatively high ecological integrity to avoid further loss of function through degradation. Once degraded, however, these wetlands offer abundant opportunities for restoration provided that the stressors causing the degradation can be reduce or eliminated. (2) When creation of a wetland becomes the selected option, design and construction specifications should be guided by data provided by reference standard wetlands of the appropriate type (e.g., HGM subclass). Using reference site data to guide the design and construction of these projects ensures that the appropriate endpoints are selected, and over time, hopefully achieved. (3) By using the same sampling protocols for monitoring sites throughout the assessment and evaluation phases, direct measures of success can be obtained, and the chances of constructing sustainable wetland projects of the desired type are greatly enhanced.

Acknowledgements

This research has been supported by a series of grants and contracts managed by the Penn State Cooperative Wetlands Center (CWC). The CWC is administered jointly by the Penn State Institutes of the

Environment and the Department of Geography of The Pennsylvania State University. The authors appreciate the contributions of L. Machung in the developmental stages of this work, and the assistance of the staff and students of the CWC. This work has benefited greatly from discussions with many employees of the U.S. Environmental Protection Agency.

References

Adamus, P.R., Brandt K., 1990. Impacts on Quality of Inland Wetlands of the United States: a Survey of Indicators, Techniques, and Application of Community-level Biomonitoring Data. U.S. Environmental Protection Agency, EPA/600/3-90/073, Environmental Research Laboratory, Corvallis, Oregon.

Adamus, P., Danielson T.J., Gonyaw, A., 2001. Indicators for Monitoring Biological Integrity of Inland, Freshwater Wetlands. A survey of North American Technical Literature (1990–2000). U.S. Environmental Protection Agency, Office of Water, EPA843-R-01, Washington, DC, 219 pp.

Bennett, R.J. 1999. Examination of Macroinvertebrate Communities and Development of an Invertebrate Community Index (ICI) for Central Pennsylvania Wetlands. M.S. Thesis. Pennsylvania State University, University Park, PA, 124 pp.

Bishel-Machung, L., Brooks, R.P., Yates, S.S., Hoover, K.L., 1996. Soil properties of reference wetlands and wetland creation projects in Pennsylvania. Wetlands 16, 532–541.

Brinson, M.M., 1993. A Hydrogeomorphic Classification for Wetlands. U.S. Army Corps of Engineers, Waterways Experiment Station, Technical Report WRP-DE-4, Washington, DC, 79pp.+app.

Brinson, M.M., Rheinhardt, R., 1996. The role of reference wetlands in assessment and mitigation. Ecol. Appl. 6, 69–76.

Brooks, R.P., 1993. Restoration and creation of wetlands. In: Dennison, M.S., Berry, J.F. (Eds.), Wetlands: Guide to science, law, and technology. Noyes Publications, Park Ridge, NJ, pp. 319–351, 439 pp.

Brooks, R.P. (Ed.), 2004. Monitoring and Assessing Pennsylvania Wetlands. Final Report for Cooperative Agreement No. X-827157-01, Between Penn State Cooperative Wetlands Center, Pennsylvania State University, University Park, PA and U.S. Environmental Protection Agency, Office of Wetlands, Oceans, and Watersheds, Washington, DC.

Brooks, R.P., Wardrop, D.H., Bishop, J.A., 2004. Assessing wetland condition on a watershed basis in the mid-Atlantic region using synoptic land cover maps. Environ. Monit. Assess. 94, 9–22.

Campbell, D.A., 1996. Comparing the Performance of Created Wetlands to Natural Reference Wetlands: a Spatial and Temporal Analysis. M.S. Thesis. Pennsylvania State University, University Park, PA, 140 pp.

Campbell, D.A., Cole, C.A., Brooks, R.P., 2002. A comparison of created and natural wetlands in Pennsylvania, USA. Wetlands Ecol. Manage. 10, 41–47.

Cole, C.A., Brooks, R.P., Wardrop, D.H., 1997a. Building a better wetland—a response to Linda Zug. Wetland J. 10, 8–11.

Cole, C.A., Brooks, R.P., Wardrop, D.H., 1997b. Wetland hydrology as a function of hydrogeomorphic (HGM) subclass. Wetlands 17, 456–467.

Cole, C.A., Brooks, R.P., 2000. A comparison of the hydrologic characteristics of natural and created mainstem floodplain wetlands in Pennsylvania. Ecol. Eng. 14, 221–231.

Dahl, T.E., 1990. Wetlands: Losses in the United States, 1780s to 1980s. U.S. Department of the Interior, Fish and Wildlife Service, FWS/OBS-79/31, Washington, DC.

Galatowitsch, S.M., van der Valk, A.E., 1994. Restoring Prairie Wetlands: an Ecological Approach. Iowa State University Press, Ames, Iowa, 246 pp.

Gray, A., Brooks, R.P., Wardrop, D.H., Perot, J.K., 1999. Pennsylvania's Adopt-a-Wetland Program Wetland Education and Monitoring Module: Student Manual. Penn State Cooperative Wetlands Center, University Park, PA, 100 pp http://www.geog.psu.edu/wetlands.

Green, E.K., Galatowitsch, S.M., 2002. Effects of *Phalaris arundinacea* and nitrate-N addition on the establishment of wetland plant communities. J. Appl. Ecol. 39, 134–144.

Hughes, R.M., Larsen, D.P., Omernik, J.M., 1986. Regional reference sites: a method for assessing stream potentials. Environ. Manage. 10, 629–635.

Karr, J.R., Chu., E.W., 1999. Restoring life in running waters. In: Better Biological Monitoring. Island Press, Washington, DC, 149 pp.

Kentula, M.E., Brooks, R.P., Gwin, S.E., Holland, C.C., Sherman, A.D., Sifneos, J.C., 1992. Wetlands. an Approach to Improving Decision Making in Wetland Restoration and Creation. Island Press, Washington, DC, 151 pp.

Kercher, S.M., Zedler, J.B., 2004. Multiple disturbances accelerate invasion of reed canary-grass (*Phalaris arundinacea* L.) in mesocosm study. Oecologia 138, 455–464.

Kusler, J.A., Kentula, M.E. (Eds.), 1990. Wetland Creation and Restoration: the Status of the Science. Island Press, Washington, DC, p. 594.

Mahaney, W.M., Wardrop, D.H., Brooks, R.P., 2004. Impacts of sedimentation and nitrogen enrichment on wetland plant community development. Plant Ecol. 175, 227–243.

Mitsch, W.J., Wilson, R.F., 1996. Improving the success of wetland creation and restoration with know-how, time, and self-design. Ecol. Appl. 6, 77–83.

National Research Council, 1992. Restoration of Aquatic Ecosystems. National Academy Press, Washington, DC, 552 pp.

National Research Council, 2001. Compensating for Wetland Losses Under the Clean Water Act. National Academy Press, Washington, DC, 322 pp.

O'Connell, T.J., Jackson, L.E., Brooks, R.P., 2000. Bird guilds as indicators of ecological condition in the central Appalachians. Ecol. Appl. 10, 1706–1721.

Rheinhardt, R.R., Rheinhardt, M.C., Brinson, M.M., Faser Jr., K.E., 1999. Application of reference data for assessing and restoring headwater ecosystems. Ecol. Restoration 7, 241–251.

Sibbing, J.M., 2003. Mitigation guidance or mitigation myth? Natl. Wetlands Newslett. 25 (1), 9–11.

Smith, R.D., Ammann, A., Bartoldus, C., Brinson, M.M., 1995. An Approach for Assessing Wetland Functions Using Hydrogeomorphic Classification, Reference Wetlands, and Functional Indices. U.S. Army Corps of Engineers, Waterways Experiment Station, Wetlands Research Program Technical Report WRP-DE-9, Washington, DC, 79 pp.

Storer, D.A., 1984. A simple high sample volumeashing procedures for determination of soil organic matter content. Commun. Soil Sci. Plant Anal. 15, 759–772.

U.S. Army Corps of Engineers, 2002. National Wetlands Mitigation Action Plan. Press release dated 26 December 2002.

U.S. General Accounting Office, 2002. U.S. Army Corps of Engineers. Scientific Panel's Assessment of Fish and Wildlife Mitigation Guidance. GAO-02-574, Washington, DC, 64pp.

Walls, R.L., Wardrop, D.H., Brooks, R.P., 2004. The impact of experimental sedimentation and flooding on the growth and germination of floodplain trees. Plant Ecology.

Wardrop, D.H., Brooks, R.P., 1998. The occurrence and impact of sedimentation in central Pennsylvania wetlands. Environ. Monit. Assess. 51, 119–130.

Wardrop, D.H., Brooks R.P., Bishel-Machung, L., Cole, C.A., Rubbo, J.M., 2004. Wetlands sampling protocol in support of hydrogeomorphic (HGM) functional assessment. In: R.P. Brooks (Ed.), Monitoring and Assessing Pennsylvania Wetlands. Final Report for Cooperative Agreement No. X-827157-01, between Penn State Cooperative Wetlands Center, Pennsylvania State University, University Park, PA and U.S. Environmental Protection Agency, Office of Wetlands, Oceans, and Watersheds, Washington, DC.

Zedler, J.B., Callaway, J.C., 1999. Tracking wetland restoration: do mitigation sites follow desired trajectories? Restoration Ecol. 7, 69–73.

Available online at www.sciencedirect.com

SCIENCE DIRECT°

ELSEVIER

Ecological Engineering 24 (2005) 341–357

ECOLOGICAL ENGINEERING

www.elsevier.com/locate/ecoleng

Construction of fens with and without hydric soils

James P. Amon [a,*], Carolyn S. Jacobson [b], Michael L. Shelley [b]

[a] *Department of Biological Sciences, Wright State University, 3640 Colonel Glenn Highway, Dayton, OH 45435, USA*
[b] *Air Force Institute of Technology, Department of Systems & Engineering Management, Graduate School of Engineering, 2950 Hobson Way, Wright-Patterson Air Force Base, OH 45433-7765, USA*

Received 17 April 2004; received in revised form 22 September 2004; accepted 1 November 2004

Abstract

The objective of this study was to show that temperate zone fens could be constructed on a surface of glacial gravel or a surface without preformed hydric soil. A reference study examined the construction of a fen on remnants of a buried fen. Each construct was replicated in four sections planted with seeds, greenhouse-grown stock or natural fen plant communities. An artesian well supplied high carbonate water through the subsurface. After 8 years, each fen contained 71 species in common, the gravel-based fen had 28 additional species and the soil-based fen had 21 additional species. On the third year, the gravel-based fen began producing calcareous peat composed of brown mosses and by the eighth year a 10 cm layer of peat covered most of the surface. The gravel-based fen produced short and less densely distributed plants than the soil-based fen but the gravel-based fen had a greater number of rare plants. While most plantings had a high degree of survival the naturally derived plugs were the most successful and the seed produced the fewest survivors. We show that, given continuous groundwater supply, it is possible to produce a high diversity fen-like environment on either substrate.
© 2005 Elsevier B.V. All rights reserved.

Keywords: Wetland; Planting; Restoration; Peat; Calcareous; Fen; Brown moss

1. Introduction

The primary goal of our experiments was to test our ability to restore fens typical of the glaciated temperate zones of the Midwestern USA and to discover the role of hydric soil in the success of that restoration. Temperate zone fens are characterized by peat or muck soils supplied by mineral rich ground water (Amon et al., 2002) that saturate the ground to the surface for much

of the year. Water does not accumulate to a significant depth in fens and most fens have a slight to moderate slope that permits the water to run away from the site. Most fens in western Ohio near the project are neutral to alkaline with significant deposits of marl (calcium carbonate), most have herbaceous cover dominated by sedges and have a high plant diversity (McCormac and Schneider, 1994). Success in meeting our goal is judged by finding that the restored site conforms to and retains the above criteria.

In our experiments, we ask whether hydric soils are required for fen restoration. Miner and Ketterling (2003) describe the accumulation of peat on gravelly

* Corresponding author. Tel.: +1 937 775 2632;
fax: +1 937 775 3320.
E-mail address: james.amon@wright.edu (J.P. Amon).

0925-8574/$ – see front matter © 2005 Elsevier B.V. All rights reserved.
doi:10.1016/j.ecoleng.2004.11.011

marl flats and it is well known that extant fens in the boreal zone developed on mineral soil bases after the last glaciation (Halsey et al., 1998). Thus, we anticipated that, given the proper hydrologic conditions, fens could be formed without using hydric soils as a base. This idea proposes that although wetlands are characterized by typical conditions of water availability, certain plants and hydric soils (Environmental Laboratory, 1987), the restoration process is capable of developing new soils. The experiment is designed to compare development of fen characteristics on both relict fen soil and on deposits of glacial outwash gravel.

Fen restoration has been attempted in numerous projects (Lamers et al., 2002) and many of these have been in locations where the peat has been harvested for commercial operations (Cooper and MacDonald, 2000; Richert et al., 2000). Some of the fen restorations have been attempted where agricultural activity adds large quantities of nitrogen and phosphorus to the ecosystem (Patzelt et al., 2001; Tallowin and Smith, 2001; Lamers et al., 2002). Few, if any, have tried to construct fens on surfaces without organic material. Since fens developed on glacial till or outwash in the post-glacial period (Andreas, 1985; Thompson et al., 1992), it should be possible to reproduce those conditions. Observations of natural fen conditions and behaviors (Wind-Mulder and Vitt, 2000; Jansen et al., 2001; Patzelt et al., 2001) have led some to propose models (Yu et al., 2001) for fens which identify key factors controlling the nature of these wetlands. The model by Yu et al. (2001) describes conditions for initiation of fens on mineral soils as having both a high and stable water table that allows peat accumulation.

Midwesten USA temperate zone fens are associated with glacial deposits of limestone-based materials contained in ancient river valleys (Andreas, 1985), a stable source of mineral rich water and geomorphology that plays an important role in fen hydrology (Thompson et al., 1992; Amon et al., 2002). Water may seep out of exposed seams of gravel to provide a water source or it may come through a break in confined aquifer that rises to the surface under pressure (Amon et al., 2002). The latter is the basis of the fen reported on in this study.

We questioned not only whether it was possible to establish fens on inorganic glacial deposits but also, whether planting them with typical fen species might rapidly produce a community of plants similar to fens native to the region. Plants found in today's typical fens may not grow well on the inorganic substrate.

2. Methods

To restore fens, we recognized that a major feature controlling their development was a water source that had sufficient hydraulic head to continuously maintain supply at the surface (Yu et al., 2001). Slope seemed important since extant fens have water that emanates from the ground, runs off easily, does not stagnate and rarely accumulates to any appreciable depth (Amon et al., 2002). The generative substrate must be porous enough to allow flow to provide saturation of the peat through most of the year and the soils that develop must also carry similar flow. The site chosen lies over a glacially buried river containing an aquifer that produces artesian wells with mineral-rich, slightly alkaline water. Fens have often been defined on vegetation basis, so we chose to use species collected from local fens to meet that need. We chose glacially deposited sand and gravel outwash from a hillside within 500 m to represent the parent substrate for our experiment. For comparison, we used hydric soils from the site of construction that appeared to be remnants of a fen buried and drained by agricultural activity. Basic construction techniques and design features were established constructing two preliminary fens and some of those details are found in Amon and Briuer (1993).

Our design was based on using or developing the appropriate hydrology, soils and vegetation for a fen. Two hydrogeologic settings are possible in fens (Amon et al., 2002). One setting with little hydrologic head that occurs on hillslopes, and the other is a mound fen that typically forms where water under pressure comes to the surface. We chose to model the mound fen because borings at the site revealed water under pressure and water in the well rose to over a meter above the planned fen surface. Fen soils were found on the site buried under agricultural non-hydric soil. That soil indicated that hydrology had once been present on site and although burial had somewhat modified the soil we chose it to represent fen soils in the experiments. Plants and seed for populating the experimental area were found nearby in natural, unaltered fens. The 15.5 m × 32.0 m site chosen was apparently once wetland but had been covered by non-hydric overwash

and intentionally by clay-rich fill. Thirty to forty-five centimetres of this overburden was over an apparently hydric, black (Munsell 2.5Y 2/0), organic soil (11.5% organic) that had high levels of clay. The overburden was removed to a disposal site and the underlying hydric soil was stockpiled on site. Under the organic soil there was an intermittent and thin (up to 15 cm) layer of clay that may have been deposited after it filtered through the organic layer. The clay was removed exposing gravel rich till that immediately produced seepage of groundwater. The near surface groundwater was found to fluctuate broadly during the year and frequently dropped below the finished surface of the fen. The primary source of water for the fen was from a 2-in. artesian well screened at 17–21 m (55–70 ft) below surface.

To assure equal water supply to the two test areas, the water supply was split to two lines controlled by brass gate valves. All pipes were made of PVC. One water line was piped to the center of each of the two sides of the site. On each side the pipes emptied into level 30 m long horizontal sections of well screen in trenches in the gravel till (Fig. 1). Before burial the flow from the screened pipe was noted to be visually equivalent along its entire length. The well screen was protected with a heavy-duty geotextile mat and then covered with about 15 cm of gravel outwash or hydric soil. The two sides of the site were separated above ground by a low mound of till to prevent mixing of hydric soils onto the gravel. Each side of the construction emptied into a separate outlet and the surface was sloped toward the exit to prevent surface accumulation of water (Fig. 1). Valves were adjusted to make the discharge into each side 8–12 L min^{-1} during spring flow.

To prepare the surface for planting, large cobbles were removed from the gravel substrate and the slope of each side was adjusted to create a fall of about 20 cm over the 30 m length to the discharge ditches. The planting area was laid out into four equal areas on each substrate (eight total). Each replicated plot was divided into zones that were planted with either single species of seed, single species of greenhouse reared plants (same species as seed plots), mixtures of seed from local wetlands, plugs dug up from local wetlands or left unplanted. All of the species planted and found are noted in Section 3. The same set of species was planted as both seed and greenhouse grown plugs (5–12-month old seedlings). A diagram of the planting pattern is shown in Fig. 2. About 100–1500 viable seeds were used in each individual seed location (0.75 m^2) and 10 of each seedling species was planted in seedling plots. Some plots were further subdivided to accommodate two or three species. To ascertain survival in a community setting a seed mixture of 45 species containing roughly 4000 viable seed was scattered over an area of 7.5 m^2. The planting zones were made accessible with lengths of 25 cm wide untreated lumber.

Planting occurred in the second and third weeks of April 1995. Seeds were prepared for planting by stratification in hydric soil from the site if they were to be planted in hydric soil or in clean quartz sand for planting in the gravel till areas. Stratification was accomplished by holding the seeds in the moist soil or sand for 8–10 weeks at 1–2 °C in the dark. Greenhouse grown seedlings were derived from the same batches of seed that were planted prior to December 1994 in the greenhouse. Greenhouse grown material was initially watered with a fen sediment slurry to provide a normal microbiota for the roots. All seed and dug up material was collected in or near the Beaver Creek Wetlands (about 500 ha of fen, marsh and shrub wetlands, in Greene County, OH, USA) adjacent to the experimental site.

Above ground biomass production was measured at the end of the first growing season by randomly collecting of triplicate 0.25 m^2 samples from each of the plots planted with a mixture of seeds. The material was dried to a constant weight at 100–103 °C to determine dry weight. Iron, calcium, magnesium, ammonia, nitrate and soluble phosphate in ground water were analyzed colorometrically as in Standard Methods (APHA, 1992). Groundwater was sampled, during the growing season, from the supply well and wells placed in the seed mixture area of each of the eight plots.

Since the function of the fen plants could be reflected in its cycles of nutrients we investigated nitrogen and phosphorus related microbial activities. Potential denitrification was assayed by gas chromotography measuring anaerobic N$_2$O production using the acetylene block method (Yoshinari et al., 1977; Quale, 1994). This method measures potential denitrification in reactions augmented by added glucose and nitrate (Quale, 1994). Nitrogen fixation was measured by the acetylene reduction method adapted from Biesboer (1984).

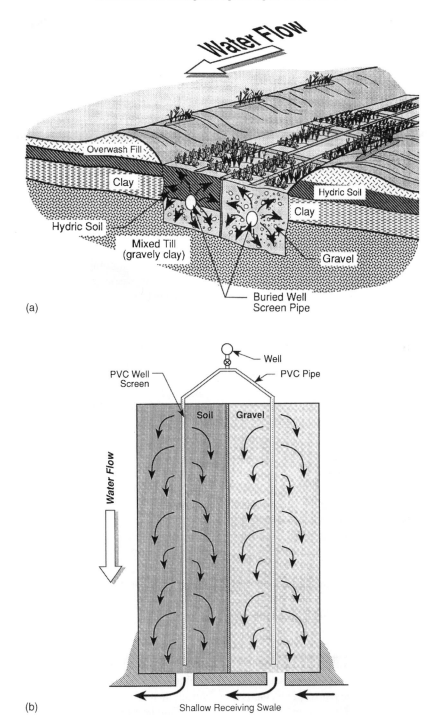

(a)

(b)

Fig. 1. (a) Perforated pipes from the artesian well are buried beneath the surface of each substrate. Both the soil and the gravel were planted with four replicate plots of the same plants and plant sources. The mound on the sides prevents surface flow from entering the fen. (b) Top view of the water supply layout. Water travels through the soil and exits via the surface then flows through a separate outlet from each substrate.

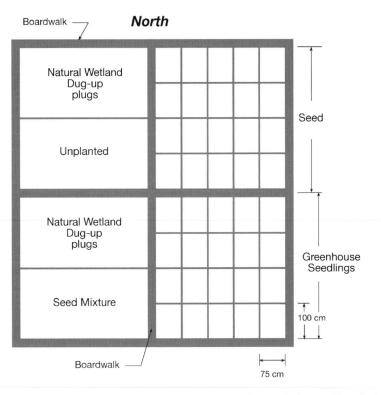

Fig. 2. Layout of the plantings. Illustration depicts pattern each of the four sections on the gravel substrate. The soil substrate is the same except that the pattern puts the seed and seedlings to the west and the remaining plantings to the east.

Mycorrhizae were investigated to determine the potential for symbiotic root fungi to enhance phosphorus uptake. Mycorrhizae were determined by staining root sections with trypan blue. Fifty microscope fields from four plants each of two species were analyzed from each substrate using the line intersect method (Brundrett, 1994). Soil samples were analyzed for bulk density, dry weights and percent organic matter, as described in Amon et al. (2002). To determine the contribution of carbonate deposits to the peat, dried cores were washed in 1N HCl until no further bubbles were detected, then rinsed in deionized water. Rinsed plant material including roots and dead material was collected on 600 μm sieves and reweighed after drying. Plants were identified by a variety of field keys and verified using Gleason and Cronquist (1991) and synonomy checked with Kartesz (1994).

To avoid sampling which could alter the outcome, many of the observations we report are necessarily qualitative in nature. Were possible we report standard deviations (S.D.) on the data gathered.

3. Results

3.1. Vegetation

All of the 33 species planted as seed produced seedlings within a few weeks. Only *Phytostegia virginiana* failed to produce survivors from greenhouse stock. Some species from greenhouse grown plugs initially appeared to die but reappeared in the second and subsequent years indicating only top growth was affected. After the first year it appeared that 9 of the 33 species planted from seed failed to survive (79% survival). In 2003, only *Angelica atropurpurea*, *Carex frankii*, *C. lupilina*, *Lobelia cardinalis*, *Mentha* sp., *P. virginiana* and *Schenoplectus tabernaemontani* remained unseen (Table 1) keeping the survival at 79%. The marsh fern, *Thelypteris palustis*, probably existed as a microscopic thallus and appeared for the first time in 2003 as an adult sporocarp. Rooting in both substrates was shallow (about 3 cm in the gravel) in the first year. During the second and third years, the gravel accumulated

Table 1
Plant distribution

Plant species	Found on soil only 2003	Found on gravel only 2003	On both soil and gravel 2003	Planted as greenhouse grown plugs	Emerging from seedbank 1995	Planted from seed	Absent 2003	Known to be in plugs from wetland	Common to Midwestern Fens (Amon et al., 2002)
Acorus calamus L.			1					1	
Agrimonia parviflora Ait.						1	1		
Agrostis gigantea Roth			1						
Agrostis stolonifera L.			1		1				
Alisma subcordatum Raf.					1	1			
Ambrosia sp. L.					1				
Angelica atropurpurea L.				1		1	1	1	
*Asclepias incarnata*L.			1	1		1		1	
Aster novae-angliae L.	1				1	1		1	1
Aster praeltus Poir.			1		1	1		1	
Aster puniceus var. firmus (Nees) Torr. & Gray		1							
*Aster puniceus*L.			1			1	1	1	
Aster vimineus Lam.						1	1		
Bidens cernua L.	1					1		1	
Bidens coronata (L.) Britt			1		1	1		1	
Bidens sp. L.		1							
Bidens vulgata Greene			1						
Bromus sp. L.		1							
Cacalia suaveolens L.	1			1				1	
Caltha palustris L.		1				1		1	1
Cardamine bulbosa (Schreb.ex Muhl.) B.S.P.			1					1	
Carex comosa Boott						1	1		
Carex cristatella Britt.			1	1		1		1	
Carex frankii Kunth				1	1	1	1	1	
Carex granularis Muhl. ex Willd.			1						
Carex hystericina Muhl.			1	1	1	1		1	1
Carex interior Bailey		1						1	
Carex laevivaginata (Kukenth.) Mackenzie		1							
Carex lasiocarpa Ehrh.			1					1	
Carex leptalea Wahlenb.		1						1	
*Carex lupilina*Muhl. ex Willd.				1			1		
Carex lurida Wahlenb.	1			1				1	
Carex pellita Muhl.			1					1	
Carex prairea Dewey ex Wood				1					
Carex sp. L.		1			1				
Carex sterilis Willd.		1						1	1
Carex stipata Muhl. ex Willd.			1	1				1	
Carex stricta Lam.			1	1		1		1	
Carex suberecta (Olney) britt.			1		1	1		1	
Carex trichocarpa Muhl. ex Willd.			1					1	
Carex utriculata Boott	1								
Carex vulpinoidea Michx.			1	1		1		1	
Chelone glabra L.			1	1		1		1	1

Table 1 (*Continued*)

Plant species	Found on soil only 2003	Found on gravel only 2003	On both soil and gravel 2003	Planted as greenhouse grown plugs	Emerging from seedbank 1995	Planted from seed	Absent 2003	Known to be in plugs from wetland	Common to Midwestern Fens (Amon et al., 2002)
Cicuta bulbifera L.			1					1	
Cicuta maculata L.			1			1		1	
Cirsium muticum Michx.						1	1		
Cyperus bipartitus Torr.			1					1	
Cyperus flavescens L.					1	1	1		
Cyperus strigosus L.			1		1			1	
Daucus carota L.	1								
Dryopteris carthusiana (Vill.) H.P.Fuchs	1								
Eleocharis erythropoda Steud.			1		1	1		1	
Eleocharis obtusa (Willd.) J.A. Schultes					1		1		
Epilobium coloratum Biehler	1							1	
Equisetum arvense L.			1					1	1
Eupatorium maculatum L.			1	1		1		1	1
Eupatorium perfoliatum L.			1	1	1	1		1	1
Euthamia graminifolia (L.) Nutt			1		1				
Filipendulua rubra (Hill) B.L. Robbins			1	1				1	
Fraxinus pensylvanica Marsh.		1							
Galium obtusum Bigelow		1						1	
Galium tinctorium (L.) Scop.			1			1		1	
Gentiana andrewsii Griseb.		1		1				1	
Helianthus giganteus L.		1							
Hypericum mutilum L.			1			1		1	
Impatiens capensis Meerb.			1					1	
Iris sp. L.			1					1	
Juncus balticus Willd.			1					1	
Juncus brachycephalis (Englem) Buch.			1	1		1		1	
Juncus dudleyi Wieg.			1		1	1		1	
Juncus effusus L.		1						1	
Juncus nodosus L.						1	1		
Juncus torreyi Coville			1	1	1			1	
Lathyrus palustris L.	1								
Leersia oryzoides (L.) Sw.	1				1			1	
Lemna sp. L.					1		1		
Leucospora multifida (Michx.) Nutt.					1				
Lobelia cardinalis L.				1		1	1		
Lobelia kalmii L.		1						1	1
Lobelia siphilitica L.			1			1		1	1
Ludwigia palustris (L.) Ell.			1		1			1	
Lycopus asper Greene		1							
Lycopus uniflorus Michx.			1						
Lycopus virginicus L.			1	1				1	1
Lysimachia hybrida Michx.						1	1		

Table 1 (*Continued*)

Plant species	Found on soil only 2003	Found on gravel only 2003	On both soil and gravel 2003	Planted as greenhouse grown plugs	Emerging from seedbank 1995	Planted from seed	Absent 2003	Known to be in plugs from wetland	Common to Midwestern Fens (Amon et al., 2002)
Lysimachia quadriflora Sims			1					1	
Lythrum alatum Pursh			1			1		1	
Melilotus officianalis (*L.*) Lam.	1				1				
Mentha piperata L.			1						
Mentha sp. L.	1			1			1		
Mimulus ringens L.			1		1	1		1	
Oxypolis rigidior (L.) Raf.			1					1	
Panicum flexile (Gattinger) Scribn.		1							
Panicum virgatum L.	1								
Pedicularis lanceolata Michx.			1	1				1	1
Pentaphylloides floribunda (Prush) A. Love	1			1				1	
Penthorum sedoides L.					1		1		
Phalaris arundinacea L.			1						
Phleum pratense L.		1							
Phytostegia purpurea (Walt.) Blake				1					
Pilea fontana (Lunell) Rydb.	1								
Polygonum amphibium L.						1	1		
Populus deltoides Bartr ex Marsh.		1			1				
Prunella vulgaris L.			1						
Pycnanthemum virginianum (L.) T.Dur & B.D. Jackson ex B.L. Robbins			1	1		1		1	1
Rorippa nasturtium-aquaticum (L.) Hayek		1							
Rosa palustris Marsh.			1					1	
Rudbeckia fulgida Ait.			1			1		1	
Rumex crispus L.			1						
Rumex orbiculatus Gray	1				1	1		1	
Sabatia angularis (*L.*) Pursh		1							
Sagittaria latifolia Willd.	1					1			
Salix amygdaloides Anderss			1						
Salix discolor Muhl.			1		1				
Salix eriocephala Michx.		1							
Salix exigua Nutt.			1		1				
Salix nigra Marsh.			1						
Salix sericea Marsh.		1							
Sanguisorba canadensis L.			1	1				1	
Schoenoplectus tabernaemontani (K.C. Gemel.) Palla				1			1		
Scirpus acutus Muhl. ex Bigelow			1					1	1
Scirpus atrovirens Willd.			1					1	1
Scirpus cyperinus (*L.*) Kunth.		1							

Table 1 (*Continued*)

Plant species	Found on soil only 2003	Found on gravel only 2003	On both soil and gravel 2003	Planted as greenhouse grown plugs	Emerging from seedbank 1995	Planted from seed	Absent 2003	Known to be in plugs from wetland	Common to Midwestern Fens (Amon et al., 2002)
Scirpus pendulus Muhl.			1	1				1	
Scirpus pungens Vahl.		1						1	
Scutellaria integrifolia L.			1					1	
Scutellaria lateriflora L.			1						
Senecio aureus L.	1					1			
Solidago ohioensis Riddell						1	1		
Solidago patula Muhl. ex Willd.	1					1		1	
Solidago riddellii Frank ex Riddell			1	1					1
Sorghastrum nutans (L.) Nash			1						
Sphenopholisa obtusa (intermedia) (Michx.) Schribn.			1					1	
Thalictrum pubescens Pursh						1	1		
Thelypteris palustris (Lawson) Fern.			1	1				1	1
Trifolium pratense L.		1							
Trifolium repens L.		1							
Triglochin maritima L.			1					1	1
Typha angustifolia L.	1								
Typha latifolia L.	1								1
Valeriana sp. L.		1	1						
Verbena hastata L.			1			1		1	
Veronica anagallis-aquatica L.					1		1		
Total	21	28	71	31	30	45	21	73	19

plant detritus and rooting developed to about 5–6 cm deep. Shallow rooted plants in the gravel greatly extended their roots horizontally, occasionally producing a visible root mass on the surface of the wet gravel. In soil, rooting was more robust and initially penetrated to at least 10 cm. By 2003, the roots in the soil substrate were abundant to at least 21 cm below surface and had entered the gravel below that. The gravel could not be quantitatively sampled. A thick cover of moss developed on the gravel substrate and formed a matrix averaging over 11 cm deep and composed of roots, moss thalli, decomposed material and calcium carbonate marl (Table 2).

By 2003, only 32 of the 45 species (71%) in the seed mixture were evident (Table 1). Twenty-six of the survivors grew on both soil and gravel and the remaining six were found only on gravel. Another set of plots, left unplanted, demonstrated the seed bank present in the

substrate and the contributions of seed from external sources. At the end of the first year, 30 species came from the unplanted soil but only eight of those were not among the various plantings (Table 1). Since first year seedlings did not produce seed, the new species that appeared had to come from either the soil seedbank or from importation. Table 1 contains the list of 119 species present in 2003 and notes those that are from locally dug natural material or planted from seed gathered. Plugs dug from local wetland sites initially produced at least 73 species of plants (Table 1). Table 1 details the plants of all origins surviving in 2003 and shows many of those that are universally found in Midwestern fens. Many of the plants seen in 2003 with unknown origin have windblown seeds but some were probably deposited by visiting animals. After 8 years of development and natural reseeding from mature plants, the distinction of planting type or location

Table 2
Soil and water characteristics

	Gravel substrate (S.D.)	Soil substrate (S.D.)
Soil development		
1995 soil depth (cm)	0	15[a]
2003 soil depth (cm)	11 (2.4)	25 (7.9)
Depth gain (8 seasons)	11 (2.4)	10 (7.9)
Percent water	84 (2.9)	50 (18.4)
1995 bulk density	NA	1.14 (0.003)
2003 bulk density (g/cc)	0.32 (0.09)	1.31 (0.60)[b]
1995 percent organic in dry weight	0	11.5 (1.82)
2003 percent organic in dry weight	26.9 (2.6)	16.8 (11.8)
1995 percent acid soluble in dry weight	NA	24 (0.4)
2003 percent acid soluble in dry weight	42.4 (2.1)	79.9 (8.9)
Water characteristics		
pH of input water	7.5 (0)	7.5 (0)
pH of water in soil/peat	7.61 (0.15)	7.69 (0.21)
Conductivity in supply water (μS/cm)	649 (6.18)	649 (6.18)
Conductivity at surface (μS/cm)	681	–[c]
Conductivity in soil/peat (μS/cm)	533 (118)	445 (100)
Free water on surface (cm)	0–3 in few places	Occurs rarely
Calcium (mg/L) in soil/peat	276 (35.1)	250 (29.0)
Calcium (mg/L) input	187 (0)	187 (0)
Magnesium (mg/L) in soil/peat	106.0 (8.5)	123.3 (8.5)
Magnesium (mg/L) input	85 (0)	85 (0)
Groundwater Ca/Mg	2.2	2.2
Peat/soil Ca/Mg	2.6	2.0
Peat/soil soluble PO_4-P (mg/L)	0.015 (0.006)	0.080 (0.010)
Peat/soil nitrate (mg/L)	0.03 (0.01)	0.03 (0.01)
Peat/soil ammonia (mg/L)	0.92 (0.14)	0.80 (0.15)
Peat/soil iron (Fe(II) mg/L)	0.59 (0.53)	0.42 (0.20)

NA implies that no soil was yet formed for the analysis.

[a] Constructed depth replicate data not available.

[b] Bulk density and dry weight increases with depth on soil but remains constant over gravel.

[c] Soil did not support free-standing water except after rain.

became indistinct and most species became well established.

The substrate seemed to have a strong influence on a few species. Of the 119 species found in 2003, 71 were present on both substrates, 28 were present only on the gravel, and 21 were present only on soil. Some species that initially germinated did not survive over the intervening 8 years. For example, *Filipendula rubra* and *Pentaphylloides floribunda* both germinated but did not grow in the gravel substrate although they survived on soil.

Some trends in the plant community were qualitatively apparent. Coverage on soil by visual estimates was about 90% in the first year. Fig. 3 compares the relative density of growth on the two substrates in the middle of the second growing season. By the end of year 2,

Fig. 3. A photograph of the experimental site taken in July of the second year. The density of growth on the soil is visibly denser than that on the gravel substrate.

the density of plant material on the soil approximated the density of the neighboring natural fens that grow on sapric soils (Linwood Muck). *Eleocharis erythropoda* became a well-established feature on both substrates in the second year, largely, by extension of rhizomes. *E. erythropoda* also seemed to spread by seed carried by animals based on the distribution away from areas planted with seed or plugs containing the species.

On the gravel substrate coverage ranged from complete in some portions to light in others. Coverage on the gravel substrate was about 50% at the end of the first year. As on the soil *E. erythropoda* spread rapidly. Where the herbaceous cover was incomplete, and water accumulated up to 3 cm deep, algae often covered the gravel substrate. The chief algae were a *Chara* sp. and *Spirogyra* sp. The filamentous green alga was most noticeable in the early spring. *Chara* grew throughout the year but the most exposed pieces decomposed after freeze damage in late winter leaving behind a considerable deposit of calcium carbonate. On both substrates *Carex hystericina*, *Scirpus atrovirens*, and *Bidens coronata* produced large populations by the second and third years as plants matured and seed production increased.

In areas not planted, seed carried in on the wind or animals populated the gravel heavily within three growing seasons. The soils had a resident seedbank that was likewise augmented by external sources. The unplanted soil area had more robust growth than the gravel but both substrates obtained good coverage similar to local fens by the third year. Plants on gravel substrate appeared similar in size and coverage to the marl flats seen in local fens. Seed was not produced by most plants until the end of the second year of growth. In year 3, seed produced at the end of year two produced a cover of seedlings most noticeable in the gravel where uncolonized space was still available. By midsummer of the third year, the growth of the soil side was robust and continued to exceed the density of the gravel side. During the first year, plants on the gravel side were denser and had a deeper green color if closer to the buried water supply pipes. Those at a distance from the pipe were somewhat chlorotic (yellow green), perhaps indicating iron or nitrogen limitation at this early stage.

At the peak of the growing season in July, during the first year, above ground biomass was slightly (*t*-test difference at $p = 0.06$) higher on the soil substrate. Dry weight was $334.7 \, \mathrm{g \, m^{-2}}$ on the soil and $203.1 \, \mathrm{g \, m^{-2}}$ on the gravel. Total phosphate per gram dry weight of biomass was also higher ($p = 0.05$) in soil ($31.7 \, \mathrm{mg \, g^{-1}}$) than in the gravel ($20.6 \, \mathrm{mg \, g^{-1}}$). Phosphate levels in the feed water appeared to be low ($<0.02 \, \mathrm{mg \, L^{-1}}$ limit of detection), and some leaf chlorosis was noted, so phosphate limitation of plants was tested by leaf and stem application of potassium phosphates (pH 7.0). There was no visual change in the height, density or color of the plants relative to adjacent controls on plants growing on either substrate indicating no phosphorus limitation. A similar experiment using NH_4NO_3 likewise produced no change indicating no nitrogen limitation. Harvest for biomass quantification was not attempted. A test of iron limitation was not attempted.

By the beginning of the 1998 growing season (third season), a clearly noticeable population of ambystilid mosses had begun to grow and some 5 cm wide mounds surrounded the base of the perennial forbs and sedges. By July of the fourth season (1999), moss growth was robust and about one-third of the previously barren areas on the gravel side were covered. During the third and fourth seasons, mosses were not yet seen on the soil side. Normal rainfall ceased in early July of 1999 and a deep drought occurred and continued through early December. While the flow of water through the system slowed it did not stop. Evaporation and a slower flow of groundwater allowed mosses to become dry at their surface but they did not die. In 2000 and 2001 (seasons 5 and 6), the mosses recovered strongly (most mounds over 35 cm wide) on the gravel side and by the fall of 2001 mosses covered a majority of spaces between the sedges and forbs except in a few locations where water stood for extended periods. In 2000 and 2001, a few mosses began to appear at the bases of some plants on the soil substrate and in 2003, nearly all of the soil spaces were populated by these mosses. In 2003, approximately three-fourth of the gravel substrate was covered by moss and the spaces not occupied with moss were covered with the algae *Spirogyra* sp., *Chara* sp. or a 2–7 cm thick layer of peaty detritus in small moss enclosed pools of water. The domininant mosses at the site in 2003 were *Bryum pseudotriquetrum*, *Campylium* sp., and *Philonotis marchica*.

Plant microbial associations were active in both substrates. Analysis showed that the gravel substrate had slightly (not significant at 0.05 level) higher trend in denitrification potential ($14–600 \, \mathrm{\mu g \, N \, g^{-1} \, day^{-1}}$) than the soil ($13–456 \, \mathrm{\mu g \, N \, g^{-1} \, day^{-1}}$), but the same residual nitrate nitrogen level in both substrates

$(0.03 \, mg \, L^{-1})$ indicated that denitrification might have a lower threshold of nitrate concentration. A natural reference fen nearby had much higher (significant at $p = 0.01$) denitrification potential $(199–840 \, \mu g \, N \, g^{-1} \, day^{-1})$. High variability of these measurements was always seen, making comparisons difficult. However, in March 1996 before plant growth, and nitrate utilization, might obscure microbial activity, denitrification occurred at a sufficient rate to lower the nitrate nitrogen levels to near detection limits $(0.01 \, mg \, L^{-1})$ in both substrates. Input levels of nitrate nitrogen from the artesian well fluctuated between 0.15 and $6.9 \, mg \, L^{-1}$. At the same time nitrogen fixation was evident. Preliminary (unpublished) work shows that nitrogen fixation activity by acetylene reduction (Quale, 1994) was well-established and fixed $10.3 \, mg \, N \, m^{-2} \, day^{-1}$ (S.D. 4.0) on soil and $6.5 \, m^{-2} \, day^{-1}$ (S.D. 5.4) on gravel by June 1996. In a natural fen site nitrogen fixation was $11.6 \, mg \, N \, m^{-2} \, day^{-1}$ (S.D. 4.2).

In the experimental area mycorrhiza were present but not necessarily more abundant than in the natural environment. Water horehound, *Lycopus americanus*, from the natural environment had 3.5% (S.D. 4.1) of its roots colonized with mycorrhiza but samples from the gravel side of the experiment were 19% (S.D. 4.8) colonized and those from the soil side were 26% (S.D. 9.4) colonized and *Mimulus ringens* was not significantly different in percent colonized on either substrates or in natural areas (natural 11.5, S.D. 3.4; gravel 14.5, S.D. 1.9; soil 15.0, S.D. 8.7).

3.2. Soil and peat formation

Accumulation of organic matter in the soil substrate was not apparent at first. Soil analysis showed initial values of organic matter in soil (loss on ignition) at about 11.5% and initial values of organic matter in gravel were undetectable (Table 2). In soil, organic matter did not change significantly over the first 3 years of monitoring (15% organic (S.D. 4.1) in 1998). By the end of 8 years the soil substrate increased in thickness and became infiltrated with a mass of roots increasing the organic matter to about 16.8% (Table 2).

The developing soil of the gravel plot was not quantitatively sampled until the eighth year. In the interim years, organic matter accumulated on the surface but initially the gravel substrate did not become well infiltrated with roots. Direct observation showed, only the top 5 cm of gravel was infiltrated with roots. On the surface a layer of decaying plant parts and algae formed a soft muddy deposit that obscured the gravel and shaded otherwise exposed roots. In some places roots lifted the sandy matrix to a higher elevation in small mounds. On the gravel substrate calcium carbonate deposits accumulated on stones, in shallow pools, and around heavy growths of *Chara* sp. Later, this marl was deposited within the peat as it accumulated (Table 2). Within the forming peat, microzones of anaerobiosis were apparent (H_2S odor and black stain under detritus) after the first season.

Formation of peat-like soils began on the gravel substrate in 1998 with the rapid expansion of the mosses originating around the bases of many plant species. By this time, a dense but shallow (5 cm) mat of roots penetrated the gravel and the peat was forming a structured setting with zones of differing redox potential (colored layers). By the end of 2003, some of those mosses were in mounds 10–15 cm thick and many mounds were merging to form a mat that covers over ninety percent of the gravel by visual estimates. In addition, dead and decaying leaves from the sedges, mosses, forbs and algae formed a mix of organic material (over 80% of dry weight after carbonates removed) even within most of the non-vegetated zones of the gravel side.

After eight seasons, soil development is still appears to be evolving. There is a large increase of soil mass over both substrates (Table 2). The increase is due to accumulation of both organic matter and deposition of carbonates. Ten centimetres of depth has been added to the soil and 11 cm of newly formed peat has been formed over gravel (Table 2). The organic matter in the soil has increased from about 11% in 1995 to over 16% in 2003 and the organic matter in gravel has increased from zero to over 26% of the total dry weight. Both soils are also accumulating significant amounts of carbonate marl. Acid extractables in the new growth over the gravel were 42% of the dry matter. In soil, acid-soluble material rose from 24% in 1995 to 79% in 2003. The bulk density of the soil 10 cm below the surface increased from over $1.14–1.31 \, g \, cc^{-1}$ (but is $0.78 \, g \, cc^{-1} \pm 0.11$S.D. in the top 10 cm) and the bulk density of the forming peat is $0.32 \, g \, cc^{-1}$. While the peat is a mass of moss evenly infiltrated with roots the distribution of roots in the soil is a gradient. Live or

recently dead roots made up $59 \pm 5.7\%$ (S.D.) of the top 3 cm of soil, $29.3 \pm 7.4\%$(S.D.) of the 3–6 cm soil layer and $11.7\% \pm 1.3$ (S.D.) of the 6–10 cm soil layer. Living moss consistently makes up the top 2 cm of the forming peat layer.

Analysis of the peat (Table 2) shows that, based on an 8-year period, about $151.3\,\mathrm{g\,m}^{-2}$ of dry peat was added per year over the gravel. Nearly half (42%) of that weight is acid soluble (carbonates). Since most of that peat was not seen in the first 3 years, the 5-year growth rate is somewhat greater. Using the 5-year period, and organic content of 26.9%, the total organic matter accumulating is about $65.1\,\mathrm{g\,m}^{-2}\,\mathrm{year}^{-1}$.

3.3. Water

Water flow from the artesian well was set at 7.5–$12\,\mathrm{L\,min}^{-1}$ in each of the two $465\,\mathrm{m}^2$ sections. Water distribution over the surface improved over the course of 3 years. During the first year, the gravel substrate had sections that would dry at the surface in hot weather. Most of these sections had larger amounts of clay mixed with the primarily sand/gravel substrate. The soil substrate saturated quickly and has stayed so continuously. By the second year, water seemed to spread more evenly into the formerly dry areas of the gravel and no drying was noted. By the third year accumulation of roots and detritus helped hold water in the formerly dry zones. Little pooling of water accumulated on the fen although spring floods caused backup of clear water on the lower half of the site for up to 3 weeks in 1996. The water depth during the flood was up to 15 cm deep. Irregularities in the surface created pools of 0–3 cm depth during normal flow, but the water moved continuously and did not appear to stagnate.

Estimated peak rate of loss of water from evaporation pan experiments in July 1996 was $16{,}443\,\mathrm{L\,day}^{-1}$. Each of the experimental sites is supplied with an estimated $9.5\,\mathrm{L\,min}^{-1}$ or $13{,}680\,\mathrm{L\,day}^{-1}$ indicating a potential water deficit during midday, but the overall 24-h loss is estimated to be about $2800\,\mathrm{L\,day}^{-1}$. Rainfall averages about 1 m per year at the site. During the warmest months of the year, surface water exiting the experimental area slowed to a trickle and stopped for only 5–8 days each year. On the soil substrate, pools did not form except during times of maximal rainfall. During the winter months, the water discharge remained high and the surface was ice bound only 10–20 days each year. During May of the first year, the unshaded gravel experienced drying at the surface. During the hot drought of summer 1999, the forming mosses dried at their surface but were not killed. During all other times water formed small and ever decreasing size, pools throughout the gravel from 1995 through 2003.

Water chemistry is typical of Midwestern temperate zone fens (Table 2). There is a slight rise in pH as the water moves through the soil and developing peat, and the dissolved calcium accumulates as the marl precipitates. During the growing season soluble phosphorus was often at or below detection limits of $0.02\,\mathrm{mg\,L}^{-1}$ both in the soil and in the input water. A few isolated readings showed soluble phosphorus as high as $0.20\,\mathrm{mg\,L}^{-1}$. Iron was quite variable and precipitated at some isolated locations of groundwater seepage. The conductivity in the water supply varied slightly but stayed between 600 and $750\,\mu\mathrm{S\,cm}^{-1}$ in the input water. Conductivity within the root zone fell as minerals precipitated. Input water was typical of temperate fens in calcium, magnesium and their ratio. There was a decrease of calcium relative to magnesium in the soil but an increase over gravel. The hydric soil accumulated more acid soluble material than the peat as it developed (Table 2).

3.4. The effect of soil contamination in a gravel plot

In May 1995, shortly after planting, a strong storm deposited a small apron of soil in the northeast corner gravel plot where seeds were planted. Within 48 h, the deposit of less than 5 mm depth was carefully scraped away leaving only a slightly visible residue of the soil. By the end of the 1995 growing season, it was noted that meter square area that had been contaminated had a more robust growth of plants than any of the adjacent portions of the same plot or any of the other three replicated gravel-based areas in the experiment. The same pattern was seen in 1996 but by 1997 the difference was not observable. Since our test for nitrogen and phosphorus limitation showed no clear indications of limitation, we tested four selected plants of each of four species in the contaminated zone for mycorrhizal colonization and compared them to those outside the zone on both soil and gravel substrate. No clear pattern emerged and no significant differences could be found.

4. Discussion and conclusions

4.1. Plants

Since composition of the plant cover varies with the substrate it appears that planting strategy should depend on both planting substrate and the plant species introduced. Nearly all plants introduced as plugs dug from the natural fens did well and plugs grown in the greenhouse had a better survival rate than seed plantings. We expected growth on the soil to be superior in both biomass production and species diversity but found that while biomass was higher in presence of the soil the diversity of plants supported was about equal on the two substrates. Numerous species were restricted to one substrate. *Lobelia kalmii* was present only on gravel and it could be that its small size and poor competitiveness with neighbors gave it an advantage there. Several of the thin-leaved *Carex* spp. (*C. interior, C. laevivaginata, C. leptalea* and *C. sterilis*) seemed to survive only the gravel substrate. All of the *Typha* spp. grew on soil and it is possible that the greater amount of weakly bound nutrient present may allow it to survive and compete with grasses and sedges. Shortly after construction we noted a short burst robust growth on soil possibly due to nutrients released during soil manipulations. In one preliminary fen, built in 1993, *Typha* sp. initially grew well on the hydric soil substrate but has been replaced by sedges, *Juncus* sp. and grasses over a span of about 10 years. Continuous flushing of the deposited nutrients in constructed fens should keep the nutrient levels low and support those plants, like sedges, that seem to grow well in low nutrient environments (Amon et al., 2002; McCormac and Schneider, 1994).

As the mosses began to dominate the surface we noted an increased number of seedlings emerging from their matrix. Mosses appear to provide moist areas with some relief from anaerobiasis typical of saturated soils. Such locations seemed to be good for seed germination and protecting seedlings that might otherwise die in periodic hot and dry weather. The apparent utility of the mosses suggests that they should be considered when plants are considered for a fen restoration.

Although biomass observations in the first year suggested superior nutrient and water availability in the soil, the gravel plantings grew well and may indicate the kind of succession that followed the retreat of glaciers from this area 12,000 years ago. Biomass determina-tions were difficult to interpret after the first year because the nature of the growth diverged. In 2003, it would be difficult for the observer to quantify the differences in the two substrates because more of the new biomass is in the developing peat and is below ground. It was obvious that plants in the hydric soil essentially filled all available space in the second year while the side with gravel till did not. The visual differences are subtle in 2003 and the appearance of the community on gravel is not much different from portions of the natural Cedar Bog fen in west Central Ohio, while the soil side is much like many parts of the Siebenthaler and Ankeney fens near to the site in Greene County, Ohio. A greater number of plants common to Midwestern fens (Amon et al., 2002) are found on the gravel substrate and both substrates support state listed (ODNAP, 2002) rare plants (*Carex lasiocarpa, Juncus balticus* and *Triglochin maritima*).

The robust growth of some plants on either substrate may be related to their colonization with mycorrhiza fungi. Many of these fungi may have been contributed by the plants and soils dug up for inclusion at the site and the established microbial symbiosis may have been responsible for the dug-up plants showing the best survival of all planting types. This conclusion is also supported by the better growth of greenhouse grown plugs that were inoculated with fen soil to introduce microbes. The fact that some plants in the restored areas were more heavily colonized by these nutritionally helpful symbionts indicates that the low nutrient groundwater may have made the associations imperative. While numerous wetland plants are known to contain mycorrhiza (Turner et al., 2000) some do not. More research is needed to determine whether these fungi should be included in restoration plantings.

Iron limitation of the communities was not tested directly, but we did observe copious iron compounds precipitated at the surface. This indicates that soluble iron II is abundant but is rapidly oxidized to and made unavailable in the near surface. During the early phases of development, the redox potential in the gravel may have favored the insoluble iron III and caused iron limitation and chlorosis seen.

4.2. Hydrology

By definition, construction of a fen requires a source of water with sufficient driving force to maintain water

at or near the surface during most of the growing season. This cannot be accomplished in many areas without the help of mechanical pumps and long-term maintenance of any mechanical device is never assured. Our site was fed by an artesian well that met the hydrological requirements without the need for mechanical devices. Given such a natural source of water, our research indicates that a fen-like wetland can be established using either hydric soils or gravelly glacial till as a substrate. The glacial deposit we used provided sufficient permeability to distribute water evenly so that the plants were constantly in contact with the water. The slightly sloped surface provided runoff that was slow without causing excessive ponding of the water. The slight irregularity of the surface after construction was complimented by the growth of plants that changed the path and retention of the water at the surface and this created numerous microhabitats that suited a great variety of plants. In one location, a plant was seen to have nearly all of its roots on the surface of the gravel under a layer of flowing water less than 1 cm in depth.

Groundwater pressures dropped during a prolonged drought but not sufficiently to damage the growth of the plants. It was obvious, during that drought, that the mosses were essential to the survival of the plantings on gravel. Construction of fens by these methods needs to consider unusual climate conditions and provide sufficient excess water capacity to maintain saturation. Since we observed a 3-year lag in establishing peat over gravel and robust root growth in soil, it seems essential that water supply during this lag period be maintained.

4.3. Soils

Site selection can be critical to the success of the construction of a fen. We built this fen where soil surveys showed a historic wetland existed (hydric soils). In addition, we looked for sites that had evidence of soils, such as muck soils, formed in seepage areas. Several agricultural fields nearby showed black soils typical of tile drained wetlands. A preliminary coring of the site identified the presence of soils with high organic content and groundwater near or potentially above the surface. In addition we knew that extensive fen systems occurred near the site. All of this information gave a high probability of finding a site to successfully construct a fen.

Excavation for our two preliminary fens demonstrated a water table well above the hydric soils and wells dug to 25 m showed the potential to bring water above the surface. Both observations suggested that seepage was historically present. The soils seen were contaminated with clay (probably washed in from fill over them) reducing their organic content, but they performed well in spite of this. Those soils produced 30 species of plants from the seed bank in the first preliminary fen constructed and that result was repeated in the present study. Seed was buried for many years and water saturated deep soils probably kept the seed viable under anaerobic conditions. While new soil formation was not evident in the first 3 years it is likely that the newly disturbed surfaces had losses of organic matter equaling the production of new material.

Although we showed that planting on gravel could produce a typical fen community, the soils that were accidentally introduced to the site provided valuable insight to the utility of soils in the restoration process. The exceptionally small amount of hydric soil introduced to the gravel provided improved growth of the plants there. In addition, the plant community growing on soils differs significantly from the community on gravel. A mosaic of soil and gravel might therefore form the basis of a richly structured fen restoration.

Formation of the new soils over gravel was rapid after the third year and although it is difficult to observe, the added volume of soil within the soil portion was about the same magnitude. The large accumulation of marl may be responsible for the composition of the plant community (Almendinger and Leete, 1998). The mosses seen are all common in local fens and seeps and many of the sedges and forbs are common in calcareous fens throughout the Midwest (Amon et al., 2002).

We were unable to find studies that reported peat accumulation on restored fens in the temperate zones in the USA. Peat accumulation rates for fens are not broadly documented but in most reports the accumulation rates reported are based on core samples, various dating techniques and the determinations of depths or carbon content (Miner and Ketterling, 2003). Peat accumulations in northern fens and bogs indicated by Gorham et al. (2003) are in the range of 15–80 g dry weight m^{-2} $year^{-1}$ and a vertical accumulation rate 0.2–1.0 mm $year^{-1}$. Thormann et al. (1999) showed peat accumulation rates of about 150–190 g m^{-2} in brown moss (fen) wetlands. Studies of more recent

depositions by Ohlson and Dahlberg (1991) report deposition rates of up to 30 mm year^{-1} in Swedish fens. Our fens forming about 65 g dry weight m^{-2} year^{-1} fall within these reported ranges. Lateral spreading of the peat much as is seen in our fens is also well known (Bauer et al., 2003) and in most northern fens it is assumed that under the proper conditions fens will eventually accumulate enough peat to isolate the surface from their mineral soil base and a transition to a bog peatland will follow (Bridgham et al., 1996; Bauer et al., 2003). Miner and Ketterling (2003) describe a situation where peat deposits are sometimes eroded indicating that the rate of peat deposition can sometimes be underestimated and can delay the transition to bogs. With a potential groundwater surface more than a meter above the restored surface, we do not anticipate transition to bog in the foreseeable future.

What we have shown is that fens can be constructed, that they have characteristics that are similar to natural fens and that the presence of hydric soils is not a necessity. We show that both hydric soils and gravelly glacial deposits can produce a highly diverse plant community and that new soils develop rapidly and become very similar to extant fen soils. The presence of denitrification, nitrogen fixation and mycorrhizal fungi indicates that microbe dependent nutrient cycles are developing. We note that plantings with stock transferred from existing fens is an excellent way to achieve success but it is also possible to do so with gathered seed and greenhouse produced seedlings. Our constructed fen is near an existing fen and it is likely that many species we recorded were imported by wind or wildlife supporting the well know notion that restoration projects may benefit from adjacent high quality habitat. Overall, the constructed fen validates the principles of fen development as outlined by Yu et al. (2001).

Acknowledgements

Rick Gardner provided external assessment of the vascular species and Robert Klips identified the mosses. Carl Friese, Monica Dorning, Terry Oroszi, Kathryn Barto, Joslin Lee and Stephen Turner assisted with mycorrhizal analysis. Teresa Carter and Bill Gruner assisted with core samples. Michelle Broyles and David Raish assisted in greenhouse work. John Blakelock assessed plant species in 2000, Songlin Cheng and Jean Li (Wright State University, Geology) provided considerable insight on the hydrogeology and hydrogeochemistry. George Hess, J. Brazelton and Christy Wendel of the (Wright State University, Chemistry) provided data on denitrification and soil organic matter. Vaughn Anderson and Randa Quale provided data on nitrogen fixation. Supported by grants from the US Army Corps of Engineers, Waterways Experiment Station and the North American Wetlands Conservation Council. This work was, in part, in fulfillment of the requirements for a masters degree (Jacobson) in Environmental Engineering at the Air Force Institute of Technology at Wright Patterson Air Force Base.

References

Almendinger, J.E., Leete, J.H., 1998. Regional and local hydrogeology of calcareous fens in the Minnesota river basin, USA. Wetlands 18, 184–202.

American Public Health Association, 1992. Standard Methods for the Examination of Water and Waste Water, 18th ed. American Public Health Association, Washington, DC, USA.

Amon, J.P., Briuer, E., 1993. Ground water is hydrology source in Ohio fen creation. Wetlands Res. Program Bull. 3, 5–8.

Amon, J.P., Thompson, C.A., Carpenter, Q.J., Miner, J., 2002. Temperate zone fens of the glaciated Midwestern USA. Wetlands 22, 301–317.

Andreas, B.K., 1985. The relationship between Ohio peatland distribution and buried river valleys. Ohio J. Sci. 85, 116–125.

Bauer, I.E., Gignac, L.D., Vitt, D.H., 2003. Development of a peatland complex in boreal western Canada: lateral site expansion and local variability in vegetation succession and long-term peat accumulation. Can. J. Bot. 81, 833–847.

Biesboer, D.D., 1984. Nitrogen fixation associated with natural and cultivated stands of Typha latifolia L. (Typhaceae). Am. J. Bot. 71, 505–511.

Bridgham, S.D., Pastor, J., Janssens, J.A., Chapin, C., Malterer, T.J., 1996. Multiple limiting gradients in peatlands: a call for a new paradigm. Wetlands 16, 45–65.

Brundrett, M.L., 1994. Clearing and staining mycorrhizal roots. In: Brundrett, M.L., Melville, L., Peterson, L. (Eds.), Practical Methods in Mycorrhizal Research. Mycologue Publications, Waterloo, Ont., Canada, pp. 51–61.

Cooper, D.J., MacDonald, L.H., 2000. Restoring the vegetation of mined peatlands in the southern Rocky Mountains of Colorado, USA. Restoration Ecol. 8, 103–111.

Environmental Laboratory, 1987. Corps of Engineers Wetlands Delineation Manual. Technical Report Y-87-1. U.S. Army Corps of Engineers Waterways Experiment station, Vicksburg, MS, USA.

Gleason, H.A., Cronquist, A., 1991. Manual of Vascular Plants of Northeastern United States and Adjacent Canada, second ed. New York Botanical Garden, Bronx, NY, USA.

Gorham, E., Janssens, J., Glaser, P.H., 2003. Rates of peat accumulation during the post-glacial period in 32 sites from Alaska to Newfoundland, with special emphasis on northern Minnesota. Can. J. Bot. 81, 429–438.

Halsey, L.A., Vitt, D.H., Bauer, I.E., 1998. Peatland initiation during the Holocene in continental western Canada. Climatic Change 40, 315–342.

Jansen, A.J.M., Eysink, F.T.W., Maas, C., 2001. Hydrological processes in a *Cirsio-Molinietum* fen meadow: implications for restoration. Ecol. Eng. 17, 3–20.

Kartesz, J.T., 1994. A Synonymized Checklist of the Vascular Flora of the United States, Canada, and Greenland, second ed. Timber Press, Portland, OR, USA.

Lamers, L.P.M., Smolders, A.J.P., Roelofs, J.G.M., 2002. The restoration of fens in the Netherlands. Hydrobiologia 478, 107–130.

McCormac, J.S., Schneider, G.J., 1994. Floristic diversity of a disturbed western Ohio fen. Rhodora 96, 327–353.

Miner, J.J., Ketterling, D.B., 2003. Dynamics of peat accumulation and marl flat formation in a calcareous fen, Midwestern United States. Wetlands 23, 950–960.

Ohio Division of Natural Areas and Preserves, 2002. Rare Native Ohio Plants: 2002–03 Status List. Ohio Department of Natural Resources, Columbus, OH, 29 pp.

Ohlson, M., Dahlberg, B., 1991. Rate of peat increment in hummock and lawn communities on swedish mires during the last 1250 years. Oikos 61, 369–378.

Patzelt, A., Wild, U., Pfadenhauer, J., 2001. Restoration of wet fen meadows by topsoil removal: vegetation development and germination biology of fen species. Restoration Ecol. 9, 127–136.

Quale, R.K. 1994. Nitrogen fixation and denitrification in a constructed fen and comparisons with a natural fen in southwestern Ohio. Masters Thesis. Wright State University, Dayton, OH, USA.

Richert, M.J., Dietrich, O., Koppisch, D., Roth, S., 2000. The influence of rewetting on vegetation development and decomposition in a degraded fen. Restoration Ecol. 8, 186–195.

Tallowin, J.R.B., Smith, R.E.N., 2001. Restoration of a Cirsio-Molinietum fen meadow on an agriculturally improved pasture. Restoration Ecol. 9, 167–178.

Thompson, C.A., Bettis III, E.A., Baker, R.G., 1992. Geology of Iowa fens. J. Iowa Acad. Sci. 99, 53–59.

Thormann, M.N., Szumigalski, A.R., Bayley, S.E., 1999. Aboveground peat and carbon accumulation potentials along a bog-fen-marsh wetland gradient in southern boreal Alberta, Canada. Wetlands 19, 305–317.

Turner, S.D., Amon, J.P., Schneble, R.M., Friese, C.F., 2000. Mycorrhizal fungi associated with plants in ground-water fed wetlands. Wetlands 20, 200–204.

Wind-Mulder, H.L., Vitt, D.H., 2000. Comparisons of water and peat chemistries of a post-harvested and undisturbed peatland with relevance to restoration. Wetlands 20, 616–628.

Yoshinari, T., Hynes, R., Knowles, R., 1977. Acetylene inhibition of nitrous oxide reduction and measurement of denitrification and nitrogen fixation in soil. Soil Biol. Biochem. 9, 177–183.

Yu, Z., Campbell, I.D., Vitt, D.H., Apps, M.J., 2001. Modelling long-term peatland dynamics. I. Concepts, review and proposed design. Ecol. Model. 145, 197–210.

Available online at www.sciencedirect.com

SCIENCE ⊘ DIRECT°

Ecological Engineering 24 (2005) 359–377

ECOLOGICAL ENGINEERING

www.elsevier.com/locate/ecoleng

Modeling the suitability of wetland restoration potential at the watershed scale

Dale White [a,b,*], Siobhan Fennessy [c]

[a] Division of Surface Water, Ohio Environmental Protection Agency, Columbus, OH 43216-1049, USA
[b] Department of Geography, The Ohio State University, USA
[c] Department of Biology, Kenyon College, Gambier, OH 43022, USA

Received 10 November 2004; received in revised form 17 January 2005; accepted 21 January 2005

Abstract

Despite the fact that landscape level processes dominate wetland ecosystem development and sustainability, restoration decisions (including those for compensatory mitigation) are typically made on a project-by-project basis. Watershed planning designed to strategically restore wetlands has the potential to provide dramatic benefits by restoring ecosystem-level processes (functions) that maintain water resource integrity. We developed a GIS-based model to predict the suitability for wetland restoration for all locations in the Cuyahoga River watershed (2107 km^2), in northeastern Ohio (U.S.A.). The model offers a useful tool to focus and set goals for wetland restoration efforts in a spatially explicit way. A two-phase approach was used: the first is to develop criteria, or environmental indicators, to identify the total population of sites suitable for wetland restoration. Locations are identified where restoration has a high likelihood of success and will be sustainable over the long term. Criteria used include hydric soils, land use, topography, stream order, and a saturation index based on slope and flow accumulation in each grid cell in the model. The second phase "filters" the total population of available sites in order to prioritize them according to their potential to contribute to water resource integrity once restored. We generated three versions of the suitability model depicting restoration potential. All versions rely on the same criteria but vary in how the factors were weighted or the hydrology criterion was calculated.
© 2005 Elsevier B.V. All rights reserved.

Keywords: Wetland restoration; Multi-criteria evaluation theory; Landscape; Northeastern Ohio

1. Introduction

The need to implement and improve wetland restoration on a watershed (including in mitigation

programs) basis was recently called for by the National Research Council (NRC, 2001). In restoration efforts, limited attention has been paid to the ways in which the spatial configuration of wetlands determines their contribution to the quality of the watershed (NRC, 2001). Restoration projects tend to be site specific, with little attention paid to processes operating at larger scales (Kershner, 1997). DeLaney (1995)

* Corresponding author.
  E-mail addresses: dale.white@epa.state.oh.us (D. White), fennessym@kenyon.edu (S. Fennessy).

0925-8574/$ – see front matter © 2005 Elsevier B.V. All rights reserved.
doi:10.1016/j.ecoleng.2005.01.012

recognized the importance of wetlands in an agricultural landscape and viewed them as integral to agricultural best-management practices at the watershed scale. He noted that if watersheds were comprised of 5–10% wetlands, they could provide a 50% reduction in peak flood period compared to watersheds without wetlands. Significantly better water quality exists in those watersheds where wetlands are incorporated into the landscape (Johnston et al., 1990; Kadlec and Knight, 1996; Day et al., 2003). A watershed planning approach designed to strategically restore wetlands has the potential to provide dramatic benefits by restoring ecosystem-level processes (functions) that maintain water resource integrity.

Selecting sites for sustainable long-term wetland restoration projects requires consideration of hydrology, geomorphology, soils, topographical variability, and surrounding land use (Bedford, 1996; Russell et al., 1997; O'Neill et al., 1997). Site selection requires an understanding of the links between wetlands and their landscapes in order to identify the hydrogeological settings that support wetlands, or what Bedford (1996) calls the "templates for wetland development". The use of a geographic information system (GIS) provides a means to analyze landscape variables on a watershed basis, and provides a means to prioritize restoration activities (e.g., Nehlsen, 1997; Russell et al., 1997). The size of the watershed, the flow processes that drive wetland functions, and the many characteristics that influence wetland biological and biogeochemical characteristics, make it advantageous to automate these procedures using GIS.

The purpose of this project was to develop a watershed level, site-suitability model using a GIS to assess the potential for wetland restoration in the Cuyahoga River watershed (CRW) in northern Ohio, U.S.A. Our goal was to identify the total population of sites suitable for restoration, then prioritize those sites according to the likelihood of restoration success and the relative ability of a restoration project at that location to contribute to downstream water quality. Existing wetlands were identified and integrated with the proposed restoration locations to maximize the potential for non-point source pollution control and increase available wetland habitat. We used the following approach:

(1) Refine criteria identified previously that will subsequently identify the total population of sites suit-

able for wetland restoration. Criteria include soil properties, land use and cover type, topographic variables, riparian zone proximity, and water quality assessments.

(2) Construct a site-suitability methodology based on multi-criteria evaluation theory in a GIS framework to incorporate the above criteria, and identify which areas have the highest likelihood of sustainability. Our ultimate goal is the enhancement of water resource integrity, defined here as the ability of a lotic system to meet Federal (U.S.A.) Clean Water Act goals for the support of aquatic life.

2. Methods of analysis

2.1. Study area

The CRW, situated in the Erie-Ontario Lake Plain ecoregion (Omernik, 1987) of northeastern Ohio, drains 2107 km^2 and includes 1965 km of perennial rivers and streams. The mouth of the watershed is within the Cleveland inner-urban zone on Lake Erie. The basin contains parts of three major physiographic provinces—the glaciated Allegheny Plateau, the till plains, and the lake plains (Ohio Environmental Protection Agency, 1994). Land use patterns vary greatly from its upper basin, which is primarily agricultural, to the lower basin, which is among the most densely populated and industrialized urban areas in Ohio. Land use in the lower basin has been highly modified by human development, and the rivers and streams situated here are influenced by the stressors associated with this land use type (for example urban and construction runoff, and hydrological modification).

Portions of the Cuyahoga River are a designated "State Scenic River" and several stream segments within the basin have been designated as high quality "State Resource Water". However, the Cuyahoga River from the Ohio Edison Dam pool to the mouth at Lake Erie (approximately 14 m upstream), and a segment of near-shore Lake Erie have been identified as an *Area of Concern* by the International Joint Commission (IJC). As required by the IJC, a *Remedial Action Plan* (RAP) was developed to improve environmental conditions. Re-establishing wetlands was identified as a priority to help alleviate flooding, reduce sediment loads, and reduce bacteria loadings from failing septic systems

(Cuyahoga River Remedial Action Plan Coordinating Committee, 1992).

2.2. Suitability modeling approach

We used a modeling approach designed to identify the spatial distribution of sites most suited to wetland restoration. The suitability modeling techniques used here are defined under the general class of *multi-criteria evaluation* (MCE) (Carver, 1991a, 1991b). MCE has its origin in planning and policy analysis. We employed the simplest of these techniques called *linear-weighted summation* or simply weighted summation. Full details of the modeling approach used can be found in Ohio Environmental Protection Agency (1998).

In developing this model, we adopt the naming convention described by Voogd (1983) employing the terms—*criterion score* and *effectiveness matrix*. Criterion scores (CS) indicate the variables that are known to influence wetland restoration. Each grid cell in the study area is assigned a score for each criterion. The CS are standardized and stored in an effectiveness matrix (**Ef**). We subsequently produced the **Ef** at a 25 m grid cell resolution.

Once the **Ef** is known, information is needed about the relative importance (weight) of each criterion. As specified by this methodology, experts provided a ranking of which criteria should be prioritized in restoration projects. We employed *paired comparisons* for determining priorities. Each individual was asked to complete a pairwise comparison matrix according to the scale identified in Table 1 (Saaty, 1977; see Ohio Environmental Protection Agency, 1998 for details on this procedure).

The individual responses are summarized in a *priority matrix* (**w**). The **Ef** matrix is linked with the **w** matrix by the weighted summation technique to generate a measure of restoration potential. In matrix terms, this entails creating an appraisal matrix (**s**), then scaling by a factor of 100 to indicate restoration potential, RP, such that:

$$E f' \cdot w = s \cdot 100 = \text{RP} \tag{1}$$

where RP is the restoration potential for a given grid cell. **Ef** is transformed (to **Ef'**) so that it is conformable with **w** in matrix multiplication.

Table 1
The nine-point paired comparison scale developed by Saaty (1977) to assign weights to criteria

Less important	
Extremely	1/9
Very strongly	1/7
Strongly	1/5
Moderately	1/3
Equally	1
More important	
Moderately	3
Strongly	5
Very strongly	7
Extremely	9

For example, if an individual felt that soil saturation was "very strongly" more important than overland flow distance, then one would enter 7 on this scale. If the inverse were the case (i.e., flow distance was "very strongly" more important than soil saturation), then one would enter 1/7.

Important assumptions behind the weighted summation technique include that the criterion weights are described using a quantitative measurement scale; the effectiveness scores are determined using a ratio scale, and the aggregation of the information (by GIS layer) takes place by addition (Voogd, 1983).

2.3. Criteria for wetland restoration

We selected criteria for use in the model on two spatial scales. The first scale consisted of physical parameters that best define wetland properties or *form*. These parameters are considered *local* in that they define the characteristics of wetland ecosystems and include the hydrologic regime, vegetative character, soil character, and topography. The second scale consisted of those parameters that best characterize wetland *function*. These parameters are considered *neighborhood* (or *landscape*) parameters including overland flow distance (i.e., distance from each square grid cell in the model to the nearest perennial stream channel), the extent to which aquatic life-use standards are met in adjacent streams, and the Strahler stream order. Neighborhood parameters provide the landscape context for a given wetland and are used to estimate its ability to contribute to downstream water quality.

Each of the criteria (whether local or neighborhood), were scaled as a "constraint" or a "factor" in the model. Constraints are binary variables and are used to represent the inclusion (score = 1) or exclusion (score = 0) of

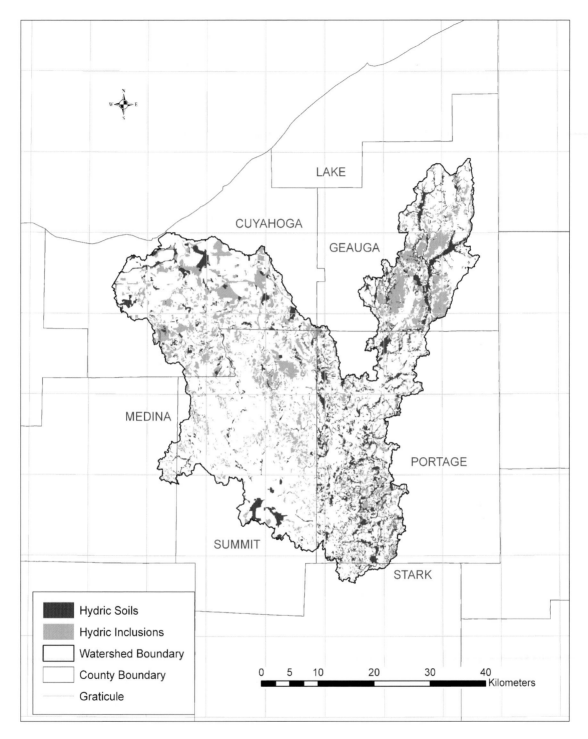

Fig. 1. Map of hydric soils and soils with hydric inclusions in the Cuyahoga River watershed (CRW). For the soil series in the CRW, the percent of hydric inclusions ranged from 2 to 15, with most values at 5%.

Table 2
Landsat Thematic Mapper (TM) land use/cover classifications as defined for use in the analysis

Land use/cover class	Definition	Criterion score
Urban	Open impervious surfaces: roads, buildings, parking lots and similar hard surface areas which are not obstructed from aerial view by tree cover	0
Agriculture/open urban	Cropland and pasture: parks, golf courses, lawns and similar grassy areas not obstructed from view by tree cover	5
Shrub	Young, sparse, woody vegetation: typically areas of scattered young tree saplings	5
Wooded	Deciduous and coniferous forest	1
Water	All open water areas including lakes, reservoirs, and wide rivers	0
Wetland	Includes wetlands identified from 1994 Landsat TM data as well as from the 1987 Ohio Wetlands Inventory (Yi et al., 1994)	5
Barren/bright surface[a]	Strip mines, quarries, sand and gravel pits, and beaches	5

Criterion scores are also provided where 0 means a cell is excluded from consideration (not suitable); 1 reflects low suitability; and 5 reflects high suitability.

[a] Many of the urban features identified in this inventory are constructed from materials obtained from the barren features. Because of this, there will on occasion be urban areas identified as barren as well as barren areas identified as urban.

a site (grid cell). We employed two constraints in this investigation—whether or not hydric soils were present in a cell (site included), and land use. Sites were included if a particular soil series was classified by the Natural Resources Conservation Service as either hydric or containing hydric inclusions. Fig. 1 portrays the distribution of soils having hydric properties (i.e., constraint = 1) within the CRW. For the land use constraint, grid cells were excluded from consideration if their land use class was urban, water, or transportation (Fig. 2).

We used five additional non-binary criteria based on the ecological underpinnings of wetland form and function. These were treated as factors, scaled to indicate their appropriateness for restoration. Two criteria indicate watershed position, including Strahler stream order and overland flow length, or the distance from each grid cell to the nearest stream channel. Two other criteria relate to the suitability of a given grid cell to support a wetland including a topographic-based saturation index (to indicate the probability of wetland hydrology developing in a cell; see section on Hydrologic Modeling below) and land use/cover type (land use class was scaled as a factor *after* excluding urban, water, and transportation land uses; Table 2). The land use and land cover constraint criterion was developed from an unsupervised classification of Landsat Thematic Mapper (TM) imagery from October 1994 (the most recent available at the time of model construction; Fig. 3). This classification approximates the USGS Level I system (Anderson et al., 1976). Wooded

land area was considered of lower value (score = 1) because of the implied costs to remove upland forest cover during wetland restoration. Definitions of each of the land use/cover categories and their respective criterion score (standardized **Ef**) are presented in Table 2.

The fifth and final factor is defined by whether the subwatershed in which a wetland will be potentially restored, contains stream segments that meet water quality standards. This factor is a means to prioritize subwatersheds for restoration; watersheds with a high proportion of stream segments that do not meet standards have a higher potential to benefit from wetland restoration. We quantified this as the proportion of stream miles in a given subwatershed that meet, at least partially, Ohio stream water quality standards. We developed the *percent difference of attainment* (PDA) as

$$PDA = [(length_{full} + length_{threat})$$
$$- (length_{partial} + length_{not})]/length_{assessed}$$

(2)

where the subscripts full, threat, partial, and not refer to full attainment, threatened, partial attainment, and non-attainment of water quality standards in each subwatershed, respectively, and $length_{assessed}$ refers to the length of the stream channel that was sampled (Fig. 4). Based on this formulation, values of PDA above zero are considered to meet aquatic life-use standards because there is a larger proportion of stream length meeting the standard. Similarly, values of PDA below zero

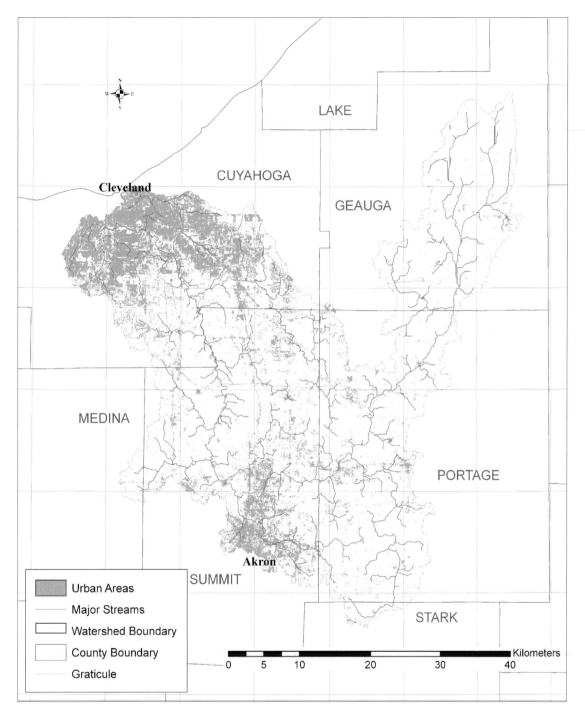

Fig. 2. Map of urban land use in the CRW showing the large land areas covered by Cleveland and Akron.

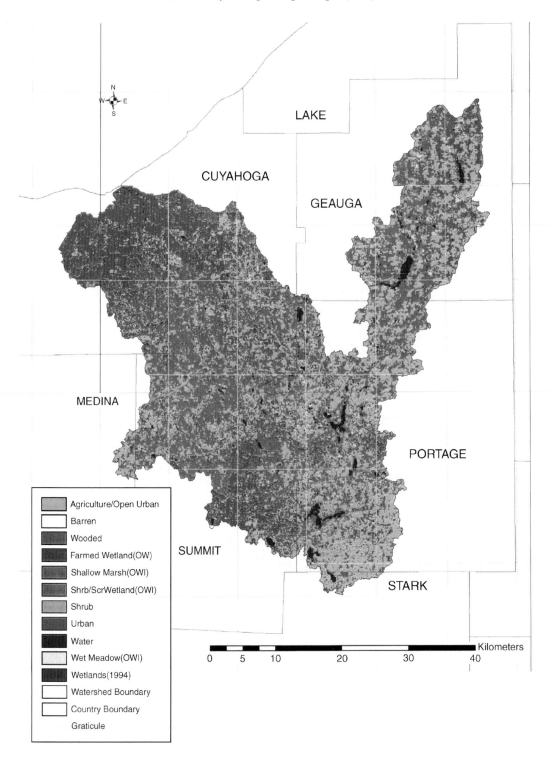

Fig. 3. Map of land use land cover in the CRW. Wetland areas are as defined in the Ohio Wetlands Inventory (OWI).

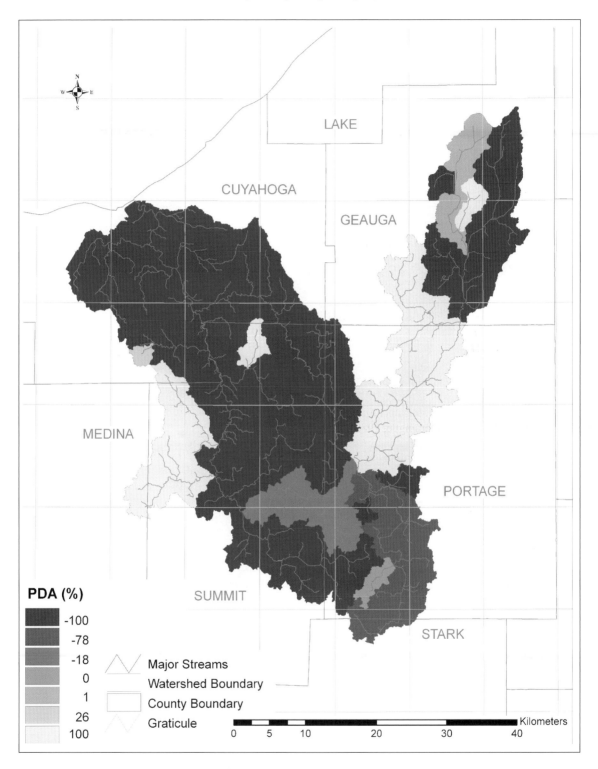

do not meet standards. Once the PDA is calculated for a segment, the downstream end of the segment is used as a subwatershed outlet and the upstream drainage area is calculated for that segment. Every grid cell in that subwatershed is then assigned the same PDA value as its outlet segment. This value is then used as a factor criterion in the weighted summation technique.

2.4. Hydrologic modeling

The topographic-based saturation index used in the suitability model was a function of the surface hydrology of the watershed. Groundwater undoubtedly influences wetland hydrology in this watershed, but because of the problems inherent in obtaining groundwater data and incorporating subsurface flows into the model, all hydrological modeling was done based on surface hydrology. A first step was to develop a digital elevation model (DEM) for the CRW. The DEM was created by interpolating elevation values from contour line data obtained from 1:24,000-scale digital line graphs (DLGs). A drainage infrastructure was then developed from the digital elevation model.

Overland flow direction was determined by searching for the direction of steepest descent from each grid cell in the DEM (Jensen and Domingue, 1988). Using flow direction, flow accumulation to any cell can easily be calculated. A recursive algorithm traces the flow paths leading to each cell in the flow direction grid and assigns a number equal to the total number of upstream cells that flow into it. When multiplied by the area of each cell, the flow accumulation value represents the contributing area for each cell in the grid. This value, along with slope is used to calculate the *topographic saturation index* (discussed below), one of the criteria used to estimate restoration potential, RP.

The drainage network and flow direction grids were used to calculate the Strahler stream order, one of two stream criteria used for calculations of RP. The contributing area for each stream segment was then delineated so that each cell in the study area was assigned a stream order value. This criterion distinguishes potential restoration sites within the watershed, from higher order to headwater stream systems.

The overland flow length from each grid cell to the nearest stream channel was our second stream-based criterion. Flow length is calculated by tracing the flow path downslope from each cell and summing the distance from the center of each cell to the center of the next down-stream cell (Fig. 5). The resulting flow length represents the hydrologic distance (along the flow path) to the nearest stream channel, rather than the total length to the bottom of the watershed. The flow length criterion reflects the overland flow distance from a potential restoration site to an existing stream channel; any water quality benefits of a restored wetland are assumed to decline with increasing distance to the stream network. Thus riparian areas are weighted as having more potential water quality benefits as sites for restoration.

The topographic saturation index indicates potential soil saturation due to surface water runoff based on upslope contributing area and slope. It is a component of the TOPMODEL hydrologic model (Beven and Kirkby, 1979). Essentially, flat locations with large contributing areas are given the greatest weight because of the concentration of runoff and low drainage potential in these areas. The slope of the terrain was calculated directly from the DEM by examining the change in elevation from cell to cell. We then used slope, along with flow accumulation, to calculate the topographic saturation index, our third factor criterion (Beven and Kirkby, 1979):

$$SI = \ln(\alpha/\tan \beta) \tag{3}$$

where SI is a unitless value representing the saturation index, α represents the upslope drainage area, and β the local slope (in $^\circ$). In addition, we developed an alternative saturation index that incorporates soil permeability (see Section 3). The alternative index provides an indication that, on balance, more water will enter the cell (via the contributing area) than will leave it through processes such as drainage and evapotranspiration, thus supporting wetland hydrology.

Fig. 4. Map of subwatersheds showing their score for the "percent difference in attainment" (PDA) in terms of meeting (attaining) water quality standards in Ohio for the protection of aquatic life. Positive values of PDA are considered to meet aquatic life-use standards since a larger proportion of stream length meets the standard. Negative values of PDA do not meet standards. *Note*: Values of PDA are multiplied by 100 for presentation purposes only.

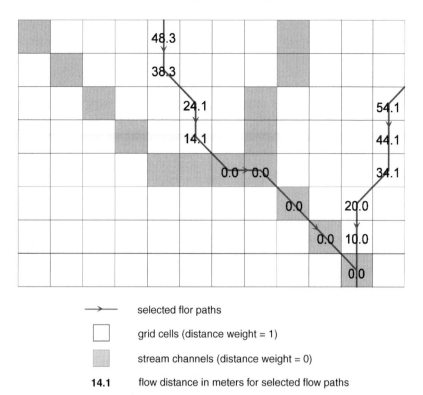

Fig. 5. An illustration of how hydrologic distance to the nearest stream channel is calculated. Flow paths are traced downslope from each cell and the distance from the center of one cell to the center of the next down-stream cell is calculated.

Table 3
Summary of data sources used to derive criteria

Data source	Derivation	Criteria derived	Resolution/accuracy
Digital elevation model (DEM)	Interpolation from 1:24,000-scale digital line graph hypsography (1998 edition) using ArcInfo® Topogrid with corrections made for stream direction	Saturation index (upstream area; local slope); stream order (stream channel depiction; watershed contribution); overland flow length (overland flow direction; stream channel depiction)	Cell size = 10 m (at generation, then aggregated to 25 m at **Ef**)
Land use and land cover	Unsupervised classification of 1994 Landsat Thematic Mapper (TM)	Land use (binary criteria); land use (factor criteria)	Cell size = 25 m; positional accuracy: 15 m
Soil attributes and location	Attributes taken from MUIR[a] (NRCS); spatial information (soil map unit polygons) derived from Ohio Department of Natural Resources (1973); date of compilation ranged from 1974 through 1982. Source map is NRCS county soil survey	Saturation index (soil permeability); hydric soil properties (binary)	Map scale = 1:15,840; minimum map unit = 1 ha (comparable to NRCS county soil survey); lines sampled from soil survey at 10 ft (3.048 m) interval; positional accuracy: 125 ft (38.1 m)

[a] Map unit interpretation records from the U.S. Natural Resources Conservation Service (1997) and developed by soil survey staff.

Table 4
Parameters used in standardization of factors criteria for suitability of wetland restoration in CRW

Criterion	Type[a]	Unit	Minimum	Maximum
Stream order	Cost	Ordinal (unitless)	1	7
Overland flow length	Cost	m	0.0	2893.28
Saturation index (without permeability)	Benefit	Continuous (unitless)	−2.35	29.22
Saturation index (with permeability)	Benefit	Continuous (unitless)	−9.06	24.87
Land use type	Benefit	Ordinal (unitless)	0	5
Use attainment (PDA)	Cost	Continuous (unitless)	−1	+1

[a] Benefit criteria are those where a higher criterion score implies a "better" score, whereas "cost criteria" are those where a higher score implies a "worse" score. To account for directionality in criterion scores, we applied the transformation of $1 - \mathbf{Ef}$ for "cost" criteria.

A summary of data sources and their attributes is included in Table 3 and whether each criterion has a positive or negative effect on restoration potential is identified in Table 4. Fig. 6 depicts how raw data sources (identified in Table 3) are related to factors- and binary criteria.

3. Results and discussion

Using readily available data and a GIS analysis we identified and prioritized wetland restoration sites for the entire CRW. We generated three versions of the suitability model depicting restoration potential. All ver-

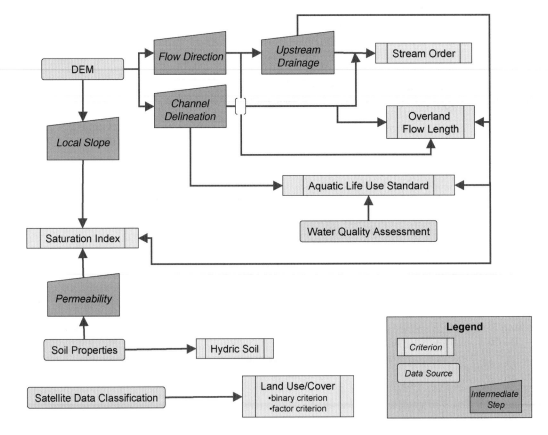

Fig. 6. Depiction of information flow from data source to criterion generation, considering intermediate steps derived from digital elevation model (DEM) and soil property data. Starting points exist at each data source (four total).

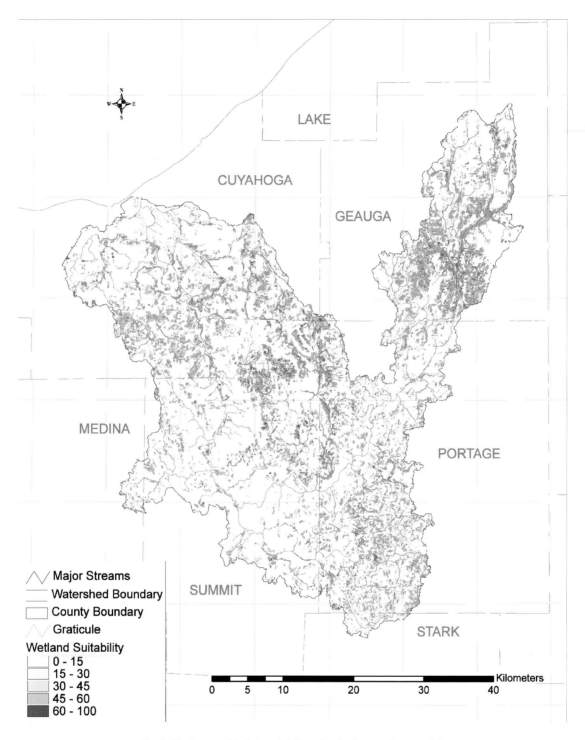

Fig. 7. The base model of site suitability of wetland restoration potential.

Table 5
Typology of model computations for wetland restoration potential in the CRW

Model type	Constraints criteria	Factors criteria	Priorities
Base	Soil property; land use	No adjustment	Average of experts based on pairwise comparison
Alternative weights variation	Soil property; land use	No adjustment	*Replacement of average set with alternative set emphasizing PDA and stream order*
Transmissivity variation	Soil property; land use	*Inclusion of soil permeability parameter in saturation index*	Average

Italics indicate major differences in the model variations.

sions rely on the two constraints—soil property and land use—and the five factors but each version varies in how the factors were weighted or how one of the factors was calculated. We identify the three computations as: (1) the *base model* (Fig. 7, (2) the *alternative weights variation* (Fig. 8 and (3) the *transmissivity variation* (Fig. 9). Differences between models are summarized in Table 5 with the distinguishing feature identified with a shaded cell. An explicit description of model variation is provided below.

The base model shows the spatial distribution of potential restoration sites ranked by categories or classes on a scale from 0–15 (no restoration potential), 15–30 (low potential), 30–45 (fair potential), 45–60 (good potential) to 60–100 (high potential; Fig. 7). Categories were established based on partitioning the range of scores, with a more inclusive top class (60–100) to reflect the stringency of the definition of this class. Palmeri and Trepel (2002) used a similar scoring method in a GIS-based system for sitting wetlands in which suitability was defined from 0 (not suitable) to 5 (most suitable). In their analysis, land suitability was estimated using a weighted average of data layers, although little detail is provided on the form and resolution of the data used.

In the base model, few areas score in the top category, with a much greater abundance of sites falling in the "good potential" class (scores of 45–60). The areas that scored in the top class tend to be located in the headwaters of subwatersheds where water quality is impaired. Spatially, the restoration sites are distributed throughout the watershed (exclusive of the urban areas). The upper northeastern peninsula of the watershed (Geauga County) shows a high density of potential restoration sites. This is due to a combination of the predominance of agricultural land use, low PDA values indicating impaired stream quality, and exten-

sive areas of hydric soils. Areas of the watershed with extensive urban land cover, including Akron (Summit County) and Cleveland (Cuyahoga County) show very few restoration sites.

The weights or priorities assigned to the different model variables can have a significant effect on the final results (Voogd, 1983). Hence, we approached the determination of weights with caution and have included variation in weights as one type of sensitivity analysis of the suitability model (the alternative weights variation).

We explored the influence of priorities on the RP, by substituting the priorities completed by pairwise comparisons (selected by the experts and then averaged) to a model with more weight being given to aquatic life-use attainment and stream order (i.e., areas with poor water quality streams and low stream orders were weighted more heavily). This was done to increase the influence that downstream water quality degradation has on the identification of restoration sites. Fig. 8 portrays the spatial distribution of RP using this alternative set of weights. The alternative weight model shows many more locations score in the highest RP class, in large part because of the weight they received from their location in subwatersheds where the streams do not meet water quality standards for aquatic life (i.e., low PDA values). As one moves downstream in the watershed, the area between Akron and Cleveland shows that nearly all of the sites identified fall into the class of high restoration potential. Akron–Cleveland is an area that is rapidly urbanizing with poor stream water quality, making the area ripe for restoration.

A second sensitivity analysis examined the inclusion of soil permeability in the topographic saturation index. Phillips (1990) describes a method of including soil transmissivity (a measure of soil permeability) in the saturation index to account for the drainage po-

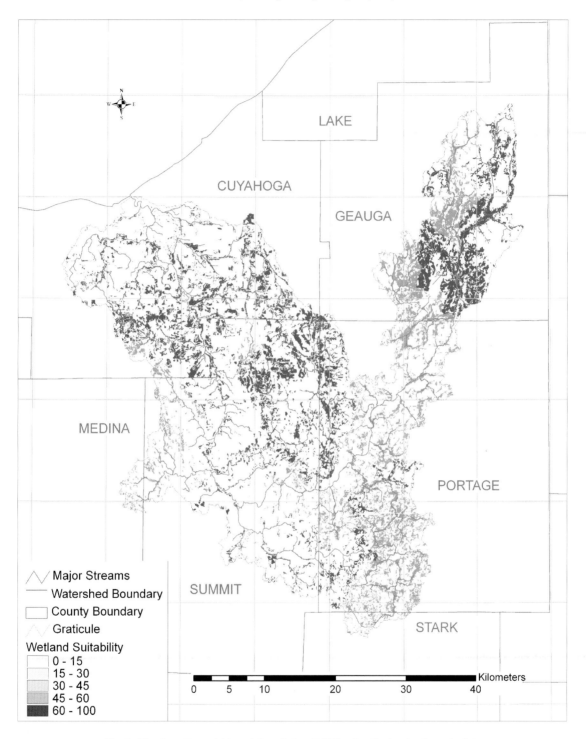

Fig. 8. The alternative weights variation of site suitability of wetland restoration potential.

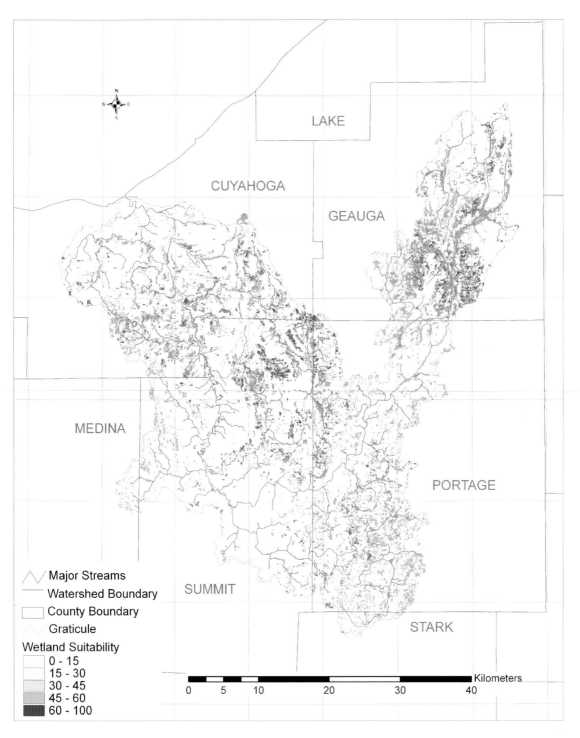

Fig. 9. The transmissivity variation of site suitability of wetland restoration potential.

tential of soils. As transmissivity increases, drainage is more rapid, and the water holding potential ("wetness") of the soil decreases. The saturation index is modified to create a relative "wetness index" (w) as follows:

$$w = \ln[\alpha/(T \tan \beta)] \tag{4}$$

where T represents soil transmissivity (note as transmissivity increases relative wetness will decrease) and α and β are defined as in Eq. (3). Transmissivity is calculated as a depth-weighted value as

$$T = \sum_{0}^{z_{\max}} (K_z D_z)/D_{\max} \tag{5}$$

where K_z equals the hydraulic conductivity (as measured by soil permeability [Phillips, 1990]) for each horizon z of the soil profile, D_z the thickness of the horizon, and $D_{\max}$ the depth from the surface to the water table or confining layer. Soil permeability data and horizon information is readily available from the MUIR database (Natural Resources Conservation Service, 1997). Fig. 9 portrays the spatial distribution of RP with the inclusion of permeability in the saturation index criterion. The spatial distribution of potential restoration sites is similar to the previous versions, and like the base model, relatively few sites scored in the highest RP category.

3.1. Consistency of model runs

We examined the similarities between the three model runs to explore if the models were substantially different from one another and, if so, to further explore these differences geographically. When comparing the quantitative differences in RP among the three models, we cannot assign differences in absolute terms because mathematically, RP = 60 from one model is not equivalent to RP = 60 from another model. The similarity in RP classes from one model to another depends on the similarity of criteria used. However, we can compare how the different models assign grid cells to the different RP classes. Overall, the percent of the total watershed area falling in the top two RP classes (good and high potential) ranges between 8.0 and 14.6% (Table 6). The base model and the transmissivity variation model assign relatively little land area to the top RP class, amounting to 0.9 and 3.1%, respectively, with much more area occurring in the second class. The alterna-

Table 6
Total and percent watershed area that exists in the top two restoration potential classes (RP values of 60–100 correspond to high potential, and 45–60 to good potential) for each model version used in this study

Model	RP class	Number of grid cells	Area (km²)	Percent of watershed area
Base	45–60	235981	149	7.1
Base	60–100	30477	19	0.9
Total			168	8.0
Alternative weights	45–60	68354	43	2.0
Alternative weights	60–100	419931	265	12.6
Total			308	14.6
Transmissivity	45–60	233852	148	7.0
Transmissivity	60–100	102153	65	3.1
Total			212	10.1
Watershed		3334855	2106	

tive weights model by comparison, assigns 12.6% land area (265 km²) to the top class, with relatively little in the second class (2.0%, or 43 km²). These values coincide with DeLaney's (1995) recommendation that watersheds with 5–10% wetland area will show substantial water quality benefits. In a model to identify riparian wetland restoration sites, Russell et al. (1997) identified only 1.87% of the land area as having medium or high restoration potential; the value is small in part because only the floodplain area was considered in the model.

Generally, all of the models have similar distributions, although the actual RP value of a grid cell may vary in the different models. We compared the coincidence of sites considered to have either a high (those scoring above the median RP) or low suitability for restoration (those scoring below the median RP) (Table 7). The model-pair with the largest agreement is the base model and the transmissivity variation whose results agree 87.4% of the time. The alternative weights version is least like the other two, agreeing approximately 73% of the time in its classification of sites. All models allocate a high distribution of RP to the northeastern peninsula portion of the CRW (Figs. 7–9) as a function of high PDA values (i.e., low water quality), the high proportion of hydric soils and inclusions (Fig. 1) and low urbanization (Fig. 2).

We also examined the spatial patterns of the three models and their relationship to the two constraints cri-

Table 7
Relative differences between model types by comparing land area within the CRW having a common restoration potential

	Base model	
	Low RP (%)	High RP (%)
Alternative weights model		
Low RP	73.1	26.6
High RP	26.9	73.4

	Base model	
	Low RP (%)	High RP (%)
Transmissivity model		
Low RP	87.4	12.7
High RP	12.6	87.3

	Transmissivity model	
	Low RP (%)	High RP (%)
Alternative weights model		
Low RP	73.3	26.9
High RP	26.7	73.1

Values represent the percent (of total) land area common to both models for values lower than the median restoration potential, low RP, and above the median restoration potential, high RP, for that particular model.

teria and the five-factor criteria. Use attainment (PDA) has a high within-index variation compared to the other criteria, thus PDA tends to dominate the distribution of RP in the three models. For example, the high values of RP in the northcentral, southeast, and northeastern peninsula portions of the watershed are influenced strongly by the high values of this criterion. Contrastingly, one-factor criterion had a very low influence on model results. The distribution of overland flow lengths is quite monotonous over the CRW and its frequency distribution is log-normal (figures not shown). Thus, overland flow length does not exert much influence on model results. In a watershed with a less articulated stream network this factor may vary more substantially and therefore help identify sites with the highest potential downstream benefits. Finally, we found the use of stream order as determined in this model problematic because it did not necessarily reflect watershed position. The calculation of Strahler stream order was segment-specific in that some low order streams connect to the mainstem rivers of the CRW. Ideally, we wanted a criterion that exhibits watershed position where greater emphasis would be placed on head-

water over lower reach portions of a watershed. Improved landscape function in headwater areas by promoting interaction of riparian and stream ecosystems through wetland restoration can improve overall watershed health.

4. Conclusions

Despite the fact that landscape level processes dominate wetland ecosystem development and sustainability, restoration decisions are typically made on a project-by-project basis. This model to predict the suitability of a site for wetland restoration offers a useful tool to focus and set goals for wetland restoration and mitigation efforts in a spatially explicit way. Implementation of the model may improve watershed condition and water resource integrity by identifying the templates for wetland development (Winter, 1988; Bedford, 1996), and may help regulatory agencies move away from the constraints that require mitigation be done "on-site" and "in-kind" (Russell et al., 1997). Our model moves substantially beyond previous approaches such as the synoptic approach to watershed assessment in which entire subwatersheds are ranked (relative to one another) based on the cumulative impacts to wetlands, but where no discrete geographic locations are identified (Abruzzese and Leibowitz, 1997). This approach is limited because it does not address the problem of locating sites individually. It also moves beyond other models that have used GIS suitability analysis to identify only one type of wetland, or wetlands limited to one location in a watershed (e.g., riparian wetlands; Russell et al., 1997). Siting potential wetland restoration projects on a watershed basis truly integrates aquatic resource protection and recovery efforts.

There are many variations that could be developed for this model. To explore those possibilities we generalize the selection of criteria into a resource and an application phase (Table 8). The resource phase considers the "raw materials' necessary to identify the population of sites available for wetland restoration. The application phase "filters" this population of available sites as a function of the intended application or goal, such as improving water resource integrity (used in this investigation), but goals could also be linked to improve terrestrial habitat connectivity, provide infor-

Table 8
Description of two project phases for selecting model criteria (both constraints and factors), and applying these to a particular water resource objective

Resource phase
Soil properties (e.g., hydric, percent organic matter, permeability)
Proximity to other wetlands (e.g., seed banks of hydrophytic vegetation)
Topographic properties (e.g., concavity, flow accumulation)
Existing land use and land cover
Existence of a appropriate hydrology (saturation index)
Land ownership (in terms of availability)
Application phase
Land ownership (in terms of cost to purchase)
Connectivity of landscape patches
Size (as a minimum area) and contiguity of adjacent land use types
Overall wetland quality desired

mation for mitigation banking, increase overall biodiversity, or minimize wetland loss by region. Using this approach we were easily able to generate alternative model versions, or scenarios, once the data compilation or resource phase had been completed. Multiple scenarios permit the use of best professional judgment in model application, allowing the evaluation of multiple alternatives (or priorities) in decision making (O'Neill et al., 1997). The flexibility inherent in this approach will allow decision makers the opportunity to explore different management options in watershed level planning, depending on local conditions and regional priorities.

Acknowledgements

We are grateful to the Region V of the U.S. Environmental Protection Agency for their financial support of our investigation. Funding was provided under Federal Grant No. CP995669-01 of the Watershed Grant Program. We thank Arnold Engelmann, currently with Danish Hydraulic Institute (Denmark), and Ming Zhang, currently with Bennett and Williams, Ohio, U.S.A.; both individuals served as technical interns for this project. Bruce Motsch and Gary Schaal of Ohio DNR Remote Sensing Program conducted the satellite land use classification. We also thank Rich Gehring, soil scientist with NRCS, and Mark Anstaett, computer programmer with Ohio DNR. We also thank the technical and editorial reviewers for improvements and suggestions to the manuscript.

References

Abruzzese, B., Leibowitz, S.G., 1997. A synoptic approach for assessing cumulative impacts to wetlands. Environ. Manage. 21, 457–475.

Anderson, J.R., Hardy, E.E., Roach, J.T., Witmer, R.E., 1976. A land use and land cover classification system for use with remote sensor data. Circular No. 671. United States Geological Survey, Washington, DC, 28 pp.

Bedford, B.A., 1996. The need to define hydrologic equivalence at the landscape scale for freshwater wetland mitigation. Ecol. Appl. 6, 57–68.

Beven, K.J., Kirkby, M.J., 1979. A physically based, variable contributing area model of basin hydrology. Hydrol. Sci. Bull. 24, 43–69.

Carver, S., 1991a. Integrating multi-criteria evaluation with geographical information systems. Int. J. Geogr. Inform. Syst. 5, 321–339.

Carver, S., 1991b. Site search and multi-criteria evaluation. Plan. Outlook 34 (1), 27–36.

Cuyahoga River Remedial Action Plan Coordinating Committee, 1992. Stage 1 Report. Public Review Summary. Impairments of Beneficial Uses and Sources of Pollution in the Cuyahoga River Area of Concern, Cleveland, OH, 45 pp.

Day Jr., J.W., Yanez, A., Mitsch, W.J., Lara-Dominguez, A., Day, J., Ko, J., Lane, R., Lindsey, J., Lomeli, D., 2003. Using ecotechnology to address water quality and wetland habitat loss in the Mississippi Basin: a hierarchical approach. Biotechnol. Adv. 22, 135–159.

DeLaney, T.A., 1995. Benefits to downstream flood attenuation and water quality as a result of constructed wetlands in agricultural landscapes. J. Soil Water Conserv. 50, 620–626.

Jensen, S.K., Domingue, J.O., 1988. Extracting topographic structure from digital elevation data for geographic information systems analysis. Photogramm. Eng. Remote Sens. 54, 1593–1600.

Johnston, C.A., Detenbeck, N.E., Niemi, G.J., 1990. The cumulative effect of wetlands on stream water quality and quantity: a landscape approach. Biogeochemistry 10, 105–141.

Kadlec, R.H., Knight, R.L., 1996. Treatment Wetlands. CRC Press/Lewis Publishers, Boca Raton, FL, 893 pp.

Kershner, J.L., 1997. Setting riparian/aquatic restoration objectives within a watershed context. Restor. Ecol. 5, 15–24.

National Research Council (NRC), 2001. Compensating for Wetland Losses under the Clean Water Act. National Academy of Sciences, Washington, DC, 322 pp.

Natural Resources Conservation Service, 1997. Soil Survey Staff. United States Department of Agriculture, National Map Unit Interpretation Record (MUIR). Accessed on 10 February 2004, available at: http://soils.usda.gov/soils/survey/nmuir/index.html.

Nehlsen, W., 1997. Prioritizing watersheds in Oregon for salmon restoration. Restor. Ecol. 5, 25–33.

Ohio Department of Natural Resources, 1973. Ohio Capability Analysis Program (OCAP). Division of Real Estate and Land Management, Columbus, OH.

Ohio Environmental Protection Agency, 1998. The Cuyahoga Watershed Demonstration Project for the Identification of Wetland Restoration Sites. Final Report to US Environmental Protection Agency, Federal Grant No. CP995669-01, 52 pp.

Ohio Environmental Protection Agency, 1994. Biological and Water Quality Study of the Cuyahoga River. Technical Report No. EAS/1992-12-11, vol. 1, Columbus, OH, 203 pp.

Omernik, J.M., 1987. Ecoregions of the conterminous United States map (scale 1:7,500,000). Ann. Assoc. Am. Geogr. 77, 118–125.

O'Neill, M.P., Schmidt, J.C., Dobrowolski, J.P., Hawkins, C.P., Neale, C.M.U., 1997. Identifying sites for riparian wetland restoration: application of a model to the Upper Arkansas River Basin. Restor. Ecol. 5, 85–102.

Palmeri, L., Trepel, M., 2002. A GIS-based score system for sitting and sizing of created or restored wetlands: two case studies. Water Resour. Manage. 16, 307–328.

Phillips, J.D., 1990. A saturation-based model of relative wetness for wetland identification. Water Resour. Bull. 26, 333–342.

Russell, G.C.P., Hawkins, O'Neill, M.P., 1997. The role of GIS in selecting sites for riparian restoration based on hydrology and land use. Restor. Ecol. 5, 56–68.

Saaty, T.L., 1977. A scaling method for priorities in hierarchical structures. J. Math. Psychol. 15, 234–281.

Voogd, H., 1983. Multi-criteria Evaluation for Urban and Regional Planning. Pion Limited, London, 367 pp.

Winter, T.C., 1988. Conceptual framework for assessment of cumulative impacts on the hydrology of non-tidal wetlands. Environ. Manage. 12, 605–620.

Yi, G.-C., Risley, D., Koneff, M., Davis, C., 1994. Development of Ohio's GIS-based wetlands inventory. J. Soil Water Conserv. 49, 24–28.

Available online at www.sciencedirect.com

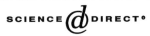

ECOLOGICAL
ENGINEERING

ELSEVIER   Ecological Engineering 24 (2005) 379–389

www.elsevier.com/locate/ecoleng

Integrating vertical and horizontal approaches for management of shallow lakes and wetlands

Thomas L. Crisman [a,] *, Chrysoula Mitraki [a], George Zalidis [b]

[a] *Howard T. Odum Center for Wetlands, Department of Environmental Engineering Sciences, P.O. Box 116350,
University of Florida, Gainesville, FL 32611, USA*
[b] *Department of Agronomy, Aristotle University of Thessaloniki, Thessaloniki, Greece*

Received 27 April 2004; received in revised form 16 December 2004; accepted 21 January 2005

Abstract

Most lake restoration/rehabilitation schemes are biased toward vertical lake management practices generally applicable to deep lakes. Unfortunately, most schemes fail to or inadequately consider their actions within the context of horizontal lake management, an especially critical component when considering shallow lakes. Two Greek lakes, phytoplankton-dominated Koronia and macrophyte-dominated Chimaditida, are used to illustrate the importance of integrating vertical and horizontal considerations in the management of shallow lakes experiencing pronounced water level reduction. Attempting to manage the structure and function of fringing wetlands via vertical manipulations of the water column are doomed to failure without consideration of changes in physical and chemical aspects of the "memory" (sediments, soils). Fringing wetlands must not be considered as monotypic habitats interacting with lakes in direct proportion to their aerial extent. A predominately vertical lake management approach is probably valid for systems such as Lake Koronia without a history of significant submersed or emergent macrophytes. For those lakes embedded within significant wetlands like Lake Chimaditida, however, failure to consider horizontal lake management as a significant component of the overall system rehabilitation will likely diminish its successful outcome. Finally, definitions of wetlands currently used by Ramsar and aquatic scientists based primarily on structural aspects of ecosystems need to be modified to recognize the overriding importance of aerially differentiated functional aspects within vegetated communities as well as fundamental differences between vegetated and open-water habitats.
© 2005 Elsevier B.V. All rights reserved.

Keywords: Chimaditida; Koronia; Phytoplankton-dominated systems; Lake management

1. Introduction

The world is facing a fresh water crisis. What historically has been a problem of water quality is fast becoming a double-faceted problem of quality and quantity. In many regions, centers of population growth

* Corresponding author. Tel.: +1 352 392 2424;
fax: +1 352 392 3624.
  E-mail address: tcris@eng.ufl.edu (T.L. Crisman).

0925-8574/$ – see front matter © 2005 Elsevier B.V. All rights reserved.
doi:10.1016/j.ecoleng.2005.01.006

and water resources do not overlap, leading governments to consider elaborate schemes both for water transfer across political and catchment boundaries and for over exploitation of groundwater resources. The World Summit on Sustainable Development in Johannesburg, South Africa, during 2002 recognized that improving the economic and health status of most people on earth is linked to critical shortages of fresh water resource quantity and quality and that this problem is rapidly expanding into a worldwide concern that is independent of economic status (United Nations, 2003).

Current and projected problems with the quality and quantity of surface freshwater resources are profound for shallow lakes and wetlands of arid and semi-arid areas of Africa (Crisman et al., 2003b), Southern Balkans (Loeffler et al., 1998; Mitraki et al., 2004), Near East (Bayar et al., 1997; Beklioglu and Moss, 1996; Green et al., 1996; Hovhanissian and Gabrielyan, 2000) and Middle East (Gophen, 2000). Many of these sites are of paramount importance for migrating and overwintering bird species and have been designated as Wetlands of International Importance (Frazier, 1996).

Both the number and magnitude of environmental perturbations to these lakes and wetlands have increased dramatically in the past two decades. Many systems have displayed progressive and profound reduction in water level due to over extraction of ground water for domestic and agricultural purposes (Mitraki et al., 2004), landscape alterations (Chapman and Chapman, 2003; Crisman et al., 2003a) and climate change (Hollis and Stevenson, 1997). Such perturbations are not limited to arid and semi-arid regions. Shallow lakes and wetlands in west-central Florida, an area of moderately high rainfall, have undergone progressive water level reductions often leading to complete dessication due to overextraction of groundwater to meet domestic demands of cities surrounding Tampa Bay (Wiley, 1997).

Even slight reduction in lake depth can produce significant changes in the structure and function of shallow lakes. Responses are complex and dictated principally by basin morphology, prior trophic state, and the balance between phytoplankton and macrophyte biomass. Littoral zones of low to moderately productive lakes with well-established macrophyte communities prior to water level reduction often expand to dominate autotrophic production in response to increased light availability, while moderately productive lakes characterized by either co-dominance of phytoplankton and macrophytes or dominance by phytoplankton tend to become completely dominated by phytoplankton following water level reduction due to increased nutrient cycling via resuspension of sediments leading to decreased light penetration and increased flocculence of sediments. Pronounced reduction in lake level in phytoplankton-dominated systems can shift the overall metabolism of the system from autotrophy to heterotrophy in spite of a progressive increase of cultural eutrophication (Mitraki et al., 2004). Such shifts to complete dominance by phytoplankton or macrophytes can result in extremely stable autotrophic conditions of low habitat diversity (Scheffer, 1998).

Limnologists have long recognized that deep lakes are inherently different from shallow systems due to their pronounced physical, chemical and biological stratification and seasonally restricted water column mixing (Wetzel, 2001), but the distinction between shallow lakes and wetlands remains unclear. While the Ramsar Convention on wetlands in 1971 defined wetlands as water bodies less than 6 m depth, it is understood that lakes and rivers of greater depth are covered entirely by the intent of the convention (Ramsar Convention and Secretariat, 2004). Such a purposefully broad definition of wetlands ignores intrinsic differences in structural and functional properties among shallow systems reflecting basin morphology, water depth, and the balance between open-water and vegetation cover. Such a definition lends confusion to any attempts to differentiate among shallow lakes, deep lakes, and wetlands to recognize the importance of water depth for inherent differences in structural and functional aspects of these three systems.

Ramsar was correct in recognizing that adjacent deep and shallow systems should be considered as a single functional unit. It is inherent in such an approach that wetlands are critical for the transformation and storage of watershed physical/chemical exports (hydrology, sediments, nutrients), thereby regulating structural and functional aspects of adjacent deep water systems. Until recently, lake management considered only the linkage between point/non-point exports from uplands and responses of open-water foodwebs, and ignored the role of vegetated aquatic ecotones as a driving factor. In marked contrast, lotic ecologists have recognized the importance of horizontal linkages with the floodplain, including the nutrient spiraling (Newbold-

et al., 1981; Webster and Patten, 1979) and flood pulse (Junk et al., 1989) concepts, as driving factors of river structure and function.

This paper examines traditional approaches for lake management and proposes an alternative approach that recognizes the role of shallow ecotonal areas as regulators of open-water structure and function. It recognizes changing relationships relative to gradients of water depth and begins to address the question of minimum water level necessary for management of aquatic systems.

2. Vertical management of lakes and wetlands

Until the recent appearance of top–down perspectives employing biomanipulation to alter foodwebs directly, most investigations approached lake and wetland management using bottom–up approaches emphasizing nutrient concentrations and availability to control primary production, and hence consumers within the foodweb (Fig. 1). Such bottom–up approaches recognize the importance of source and non-point horizontal loadings of nutrients and sediments from the watershed as the ultimate control over production of aquatic ecosystems, but until very recently few studies looked at the importance of shallow water vegetated areas in the transformation (chemistry to biomass) and storage of such inputs prior to their entering open-water. Processing of watershed chemical, sediment, and hydrological signals may be via incorporation into floral and faunal biomass, short to medium term storage in biomass and detritus, and export to open-water as living biota, detritus, or dissolved nutrients and carbon. A secondary source of chemical loading, the atmosphere–water interface, has been the focus of particulate and dissolved loading mostly in ultra-oligotrophic lakes embedded within nutrient poor geologic matrices and/or influenced by pronounced acid rain (Psenner, 1994). Regardless of source, such management approaches have focused on either measure to reduce nutrients at their source or to intercept them prior to entering the aquatic ecosystem (Mitsch et al., 2001).

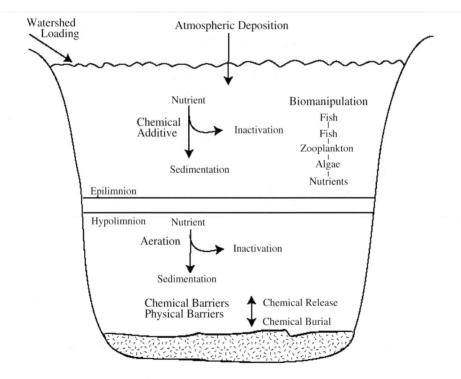

Fig. 1. Common approaches to the vertical management of lakes that focus on the water column (epilimnion and/or hypolimnion) and the sediment–water interface.

There are two general categories of in-lake management approaches: sediment–water and water column interactions (Cooke et al., 1993). Management practices focused on sediment–water interactions seek to reduce nutrient release to increase permanent storage and nutrient availability for biotic production within the water column, thereby reducing potential lag times between implementation of a management action and a positive ecosystem response. Both physical barriers of silt and clay overlying lacustrine sediments and creation of chemical barriers within or at the surface of sediments utilizing phosphorus binding compounds like alum have been effective at stopping recycling of phosphorus from relatively firm, mainly macrophyte produced sediments (Hansen et al., 2003). Similarly, chemical inactivation via addition of either solutions or powders of an aluminum salt, usually aluminum sulfate (alum) and or sodium aluminate (Reitzel et al., 2003), ferrous iron (Deppe and Benndorf, 2002), calcite (Dittrich and Koschel, 2002) or gypsum (Salonen et al., 2001) into the water column can reduce nutrient availability for primary production of phytoplankton, often forming precipitates that are eventually deposited on the lake bottom promoting long term nutrient storage in sediments.

Total water column aeration has proven effective at oxygenating both the water column and the sediment–water interface in shallow lakes characterized by extremely flocculent sediments resulting from phytoplankton domination of autotrophic production to reduce nutrient availability (Cooke et al., 1993). In lakes deep enough to develop a stable hypolimnion, however, hypolimnetic aeration is preferred, as it does not promote water column holomixis, is effective at reducing sediment nutrient release and provides an oxygenated deep water refuge for fish and invertebrates (Soltero et al., 1994).

The addition of biomanipulation to alter the structure and biomass of pelagic foodwebs (top–down) is a relatively recently alternative to traditional bottom–up management approaches that focus on reducing nutrient availability to the primary production base of the ecosystem (Carpenter and Kitchell, 1989). Most biomanipulation schemes seek to restructure fish communities to either promote a cascading of interactions among trophic levels of the foodweb leading to altered taxonomic composition and/or biomass reduction of the phytoplankton assemblage via elevated predator fish populations, direct grazing on phytoplankton by phytophagous fish or decreased nutrient cycling from sediments by elimination of the bioturbation actions from benthivorous fish such as catfish or carp. An indirect benefit of biomanipulation is often reduction in nutrient cycling via sedimentation of intact feces and degradation resistant body parts. All biomanipulation schemes seek to alter biotic structure directly to alter the function of pelagic food webs independent of controls over nutrient availability.

Thus, traditional management practices of open-water ecosystems have taken a strictly vertical perspective incorporating air–water, sediment–water and within water column interactions. Such approaches recognize the importance of horizontal loadings from the watershed, but largely ignore the importance of shallow water vegetated areas at mediating such inputs prior to entering open-water. As such, the solution to management of shallow lakes experiencing significant loss of water level has taken a vertical perspective to add sufficient water to reduce the importance of sediment–water interactions for nutrient cycling and to provide sufficient oxygen to favor fish communities capable of structuring the pelagic foodweb to support a specific management object (Crisman, 1986).

3. Horizontal management of lakes and wetlands

Shallow lakes are controlled more by system memory (sediments) and interactions with the water column than are deep lakes because of their lack of thermal stratification and increased sediment disturbance thorough wave and current action. Given that the shallower the lake, the greater the interactions with sediments and the lower water column, management of such lakes has focused primarily on vertical approaches to minimize recycling of sediment memory by increasing the depth of the water column (Cooke et al., 1993). Increased stability of the sediment–water interface favors permanent burial of nutrients, thereby reducing the lag time between implementation of a management scenario and a positive response of the aquatic system.

Reflecting basin configuration, shallow lakes tend to be surrounded by extensive littoral zones and wetlands, and are thus also strongly influenced by horizontal processes and interactions (Fig. 2). As noted earlier,

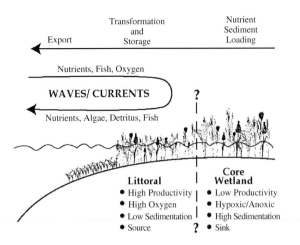

Fig. 2. Processes operating from the terrestrial–wetland and wetland–pelagic interfaces in shallow lakes emphasizing differences between the littoral zone and the core wetland.

vegetated littoral zones and fringing wetlands receive loadings of chemicals and sediments from uplands, and through transformation and storage processes, in turn regulate their subsequent export rate and form (detritus, living biomass, dissolved nutrients) to open-water. Therefore, shallow vegetated areas strongly control pelagic metabolism (Wetzel, 2001). This metabolic kidney function is short circuited, however, when loading rates exceed the growth capacity of the plant community in combination with water residence time in the wetland. Wide, densely vegetated shorelines have the greatest likelihood of having the longest residence time and achieving maximum nutrient uptake via plant and sediment processes, including denitrification, with possible medium to long-term storage in accumulating sediments.

The functional role of a fringing wetland in lake metabolism is controlled not only by its width, plant density, and productivity, but also by the extent of direct interactions at its ecotonal boundary with the pelagic area of the lake (Fig. 2). The width of the pelagic–wetland interaction zone is influenced mainly by a balance between wetland hydrological export and the counterbalancing force of waves and currents entering the wetland from the pelagic, mediated by the physical obstruction offered by plant density and growth habit (free-floating, emergent, submersed). The width of this zone can be highly variable, extending just a few meters into fringing wetlands of even large lakes characterized by strong currents and long fetch, and its

innermost boundary can be characterized by a sharp declining gradient in oxygen (Chapman et al., 2002; Rosenberger and Chapman, 1999).

While near shore portions of fringing wetlands function as nutrient sinks through plant transformations, sediment accumulation and denitrification, the wetland–pelagic interaction zone tends to be a source of nutrients and organic matter for the pelagic zone (Fig. 2). Wave and current action keep the area well oxygenated and productivity of attached algae is expected to be greater than that of interior portions of wetland beyond pelagic zone influence. Although poorly studied, it is clear that wetlands can be major exporters of fish and invertebrate biomass and detritus to the pelagic zone (Wetzel, 2001). There are also major interactions from the pelagic to the vegetated areas. Many pelagic fish species use near shore vegetated areas for breeding and exhibit a sequence of on–offshore movements at various parts of their maturation cycle reflecting both varying needs in food quality and quantity and refuge from differential predation pressure (Frankiewicz et al., 1996). While increasing biomass of submersed macrophytes can benefit populations of the predaceous invertebrate midge *Chaoborus* and fish, there is a progressive loss of abundance and population fitness of both as the percentage of the water column colonized by macrophytes increases above 50% in lakes (Crisman and Beaver, 1990).

One critical unanswered question remains regarding the horizontal extent of vegetated shallows necessary for proper management of adjoined shallow lakes. Such areas are designated littoral zones by limnologists and fringing wetlands by wetland scientists. The distinction is not one of semantics, but rather one of implied function. Historically, wetlands scientists were botanists viewing wetlands as waterward extensions of terrestrial plant communities, while limnologists viewed them as landward extensions of submersed macrophyte communities. It is time to reconcile differences in such definitions and to assign functional attributes.

It is proposed that the structurally based definition of littoral zones (Wetzel, 2001) be modified to include a functional landward boundary that delineates the limit of the wetland–pelagic interaction zone, thus recognizing its paramount importance in the metabolism of the pelagic (Fig. 2). The remainder of the emergent vegetated shallow water would constitute a core wetland area that interacts with open-water only during either

major storm events and associated waves or major rain events to promote watershed runoff. It is important to note that while shorelines of individual lakes can be structurally similar from the shore outward to open-water, this definitional modification removes the implication that structure implies function throughout. Such a distinction will become increasingly important in the management of shallow lakes surrounded by extensive wetlands when only a portion of the wetland can be expected to be saved from anthropogenic activity, especially conversion to agriculture.

Key to this ammended definition of the littoral is the ability to define precisely its horizontal inner boundary. While the core wetland tends to be an area of lower sessile algal productivity, hypoxic to anoxic conditions, and medium to long-term sediment accumulation, the wetland–pelagic interface (littoral) is an area of high-algal productivity, high-oxygen concentrations through pelagic wave and current action and a low-sediment accumulation via the action of current winnowing to open-water. Therefore, selection of parameters for assessing the boundary between these two zones should consider that the core wetland is a sink, while the littoral is a highly dynamic interaction zone and source of nutrient and carbon export to the pelagic. Parameters with great promise for defining the littoral-wetland boundary are dissolved oxygen, specific conductivity and pH for short-term (hourly–daily changes in metabolism and water movement), attached algae and benthic invertebrates for medium-term (monthly–seasonal) and sediment chemical profiles for long-term (interannual) temporal boundary position and conditions. The relative partitioning of shallow vegetated areas of lakes into littoral and core wetlands will be site specific and strongly controlled by bottom configuration; the horizontal extent, density and growth habit of the vegetation; and basin size and orientation to dominant wind direction.

4. Integrating vertical and horizonal management

Case histories from two lakes in northern Greece are used to illustrate integration of vertical and horizontal approaches in the management of shallow lakes surrounded by extensive wetlands (Chimaditida) versus those without (Koronia). Both lakes have undergone pronounced water level reduction during the past 20 years as a result of agricultural activities.

4.1. Lake Chimaditida

Lake Chimaditida (21°34′05″ longitude, 40°35′45″ latitude) is approximately 200 km west of Thessaloniki and receives water from smaller Lake Zazari via a 2 km long stream. The combined watershed for these lakes is 228 km², much of which is in crop production. Chimaditida has experienced major water-level decline in the past three decades from a maximum depth of 8 m in 1970 to approximately 1.8 m in 2001. Drainage of the extensive marsh system to the north of the lake for agricultural production began in the early 1960s and was expanded later in the decade to 22 km², when a dike 0.5–1 m high was constructed at the lake outlet to regulate and lower water level.

The major water-level reduction following drainage, diversion actions and dike construction in the 1960s and 1970s resulted in a profound reduction in the landward extent of the *Phragmites*-dominated wetland and a loss of lake pelagic area from wetland expansion into open-water. After completion of most of drainage operations in 1970 and associated loss of the most landward wetland portions, Chimaditida still had wetland and open-water areas of 91 and 946 ha, respectively. By 1997, however, the open-water area had been reduced 83% to 164 ha, while wetland area had expanded 10 times to 904 ha through invasion of previously pelagic areas of the lake.

As with a majority of Greek lakes, there are few historical data to evaluate the total ecosystem response to major environmental perturbations. Cyanobacteria blooms are common, but the limited phosphorus data (range 1250–1650 μg/L for 1986 and 1992) are insufficient to delineate the history of trophic state for the lake.

4.2. Lake Koronia

Lake Koronia (23°08′48″ longitude, 40°41′21″ latitude), approximately 15 km northeast of Thessaloniki, was once the fourth largest lake in Greece, but it recently has shrunk drastically from its 4620 ha extent in the 1980s to approximately 3440 ha as a result of lowering of lake level from >4 m to <1 m (Zalidis and Mantzavelas, 1994). The lake lies in a tectonically

active faulted basin, and its watershed (approximately 350 km^2) is drained by three creeks and a ditch. It is listed as a Wetland of International Importance by RAMSAR in recognition of its importance for nature conservation, especially birds. Although the lake had an active commercial fishery in the late 1980s, concerns were expressed for its sustainability as a result of progressive water-level decline (Economidis et al., 1988; Fotis et al., 1992). Unfortunately, the fishery totally collapsed during the 1990s.

Agriculture intensified in the basin during the 1960s and especially the 1970s accompanied by a progressive increase in the number of irrigation wells. Additional stress on groundwater resources was associated with establishment of industries at the western end of the lake in the 1970s and their expansion in the 1980s. The main point-source discharge for agricultural and industrial activities to the lake is a ditch at the western end of the lake, and heavy metal concentrations in sediments are high where the stream enters the lake (Anthemidis et al., 1997).

Lake Koronia has experienced a progressive increase in trophic state associated with decreasing water level since the early 1990s (Mitraki et al., 2004). Water-column conductivity remained at 1100–1300 μS cm^{-1} from 1977 to 1989, then increased rapidly to exceed 6500 by 1996. The most rapid increase was during 1994–1996. An identical trend is displayed by total phosphorus, with values remaining at <200 μg/L from 1977 to 1989, then increasing rapidly to peak values of >1000 μg/L by 1996. Lake Koronia displayed the second highest chlorophyll (206 mg/m^3) and lowest Secchi disk transparency (0.2 m) of 14 Macedonian lakes survey by Koussouris et al. (1992) and was clearly hypertrophic by the end of the 1980s. Conditions have only gotten worse since then as macrophytes have disappeared and the system has become completely dominated by cyanobacteria.

The period of greatest increase in trophic state coincided with the progressive reduction in lake level from at least 1986 (water depth = 4.0 m) through 2001 (water depth < 1.0 m). This decline was independent of regional rainfall trends and is attributed to overextraction of ground water resources for agricultural and industrial purposes (Mitraki et al., 2004). The Mygdonia Basin, which includes Koronia and adjacent Lake Volvi, has over 2069 wells for irrigation purposes, of which 1091 are in the vicinity of Lake Koronia

(HMEPPPW, 1996). Most of these, as well as those for industrial purposes, were drilled in the late 1970s and 1980s prior to legal curtailment of new wells in the early 1990s.

4.3. Restoration scenarios

Proposals are being formulated to increase water levels in both lakes Koronia and Chimaditida in order to enhance conservation value and reverse the effects of cultural eutrophication. While Scheffer (1998) recognized both fundamental differences in the structure and function of shallow versus deep lakes and the presence of alternative stable states for autotroph dominance (macrophytes and phytoplankton), initiation of a shift between these two states is strongly controlled by system "memory" (sediments) and hydrology.

A conceptual model is proposed for the response of phytoplankton (Koronia) and emergent-macrophyte (Chimaditida)-dominated shallow lakes to altered water level that incorporates sediment physical influences on likely ecosystem restoration success (Fig. 3). Lake Koronia was phytoplankton-dominated prior to hydrologic alterations with few submersed and emergent macrophytes. Our ongoing paleolimnological investigation of the lake suggests that sediments deposited at this time were uniform in character and somewhat flocculent. The progressive lowering of water level since at least the 1980s has established cyanobacteria as the dominant autotrophs in the lake resulting in highly flocculent sediment that is readily suspended into the shallow water column. Emergent macrophytes only sparsely colonize the rapidly retreating shoreline, and submersed macrophytes are essentially absent due to low transparency and flocculent sediments.

Although it is unlikely that any rehabilitation scheme for Lake Koronia will have sufficient water to return water depth to 1980 values, the relative dominance of phytoplankton and macrophytes will depend on both the temporal extent of sediment exposure to the atmosphere and the speed of water level change. The longer the drawdown period, the greater the extent of dewatering, compaction and oxidation of exposed sediment to affect nutrient cycling. Under this scenario, flocculent sediments would be aerially restricted to the deepest portion of the basin that is continually flooded, while sediments higher in the basin morphometric

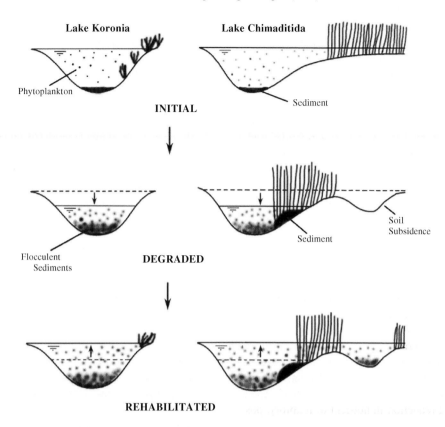

Fig. 3. Likely responses of two Greek lakes, Koronia and Chimiditida, to progressive water level reduction and hydrological rehabilitation.

profile could support rooted macrophytes upon reflooding because of development of a firm substrate.

Key in establishing a vegetated littoral zone is the speed of water-level return and the slope/configuration of the basin. The faster the return, the less likely the establishment of submersed macrophytes due to shading from the likely continuing dominance of cyanobacteria supported by sediment release of nutrients from flocculent sediments in deeper water and potential pumping of nutrients from sediments by macrophytes. Thus, it is extremely difficult to reverse phytoplankton dominance without close attention to physical and chemical characteristics of system "memory" (sediments). Thus, raising water level will likely produce little change in cyanobacteria as the dominant stable state of the lake.

Lake Chimaditida shifted to nearly complete dominance by emergent macrophytes (*Phragmites*) following profound reduction in water level (Fig. 3). Former wetland areas were farmed and the wetland fringe expanded into the pelagic to reduce its extent signif-

icantly. It appears that exposure of organic wetland soils for agricultural production has promoted soil compaction, dewatering and decomposition, thus lowering soil surface elevation over the past two decades. Similar operations in former wetlands of Louisiana and the Florida Everglades have resulted in soil subsidence approaching 3.0 cm/year (Shih et al., 1998; Trepagnier et al., 1995). Conversely, *Phragmites* expansion into formerly pelagic areas has likely increased accumulation of organic sediments as suggested by the 0.25–1.1 cm/year rates noted by Reddy et al. (1993) for a comparable emergent macrophyte, *Cladium jamaicense*, in the Florida Everglades. Pelagic areas of eutrophic Lake Chimaditida would be expected to accumulate sediments at <1.0 cm/year (Brenner et al., 1999a,b). As happened at Lake Apopka, Florida, longterm farming of former wetlands surrounding a eutrophic shallow lake can result in the sediment surface of the former wetland being meters below that of the adjacent lake bottom.

Increasing water level in Lake Chimaditida will likely push the *Phragmites* front closer to the shore as water depths exceed plant colonization and maintenance limits, but because of sediment accumulation patterns and associated shoaling, it is unlikely that the front will return to its 1970 position. In addition, reflooded former wetland areas are likely to develop into a mosaic of open and vegetated areas reflecting differential sediment loss during the drawdown period, and there is likely to be significant flooding associated phosphorus release at least in the short-term (Calzada-Bujak et al., 2001).

5. Implications for lake wetland management

Most lake restoration/rehabilitation schemes are biased toward vertical lake management. Lakes are considered from a pelagic perspective, whereby increasing water level will reduce the manifestation of cultural eutrophication by reducing nutrient availability in the water column through reduced physical sediment-resuspension and possible trapping in a restored hypolimnion (provided the lake is deep enough). Even without a reduction in nutrient availability, possibly higher oxygen concentrations under higher water regimes may favor biomanipulation of pelagic and benthic components of the foodweb to enhance grazing on excess autotrophic production (Carpenter and Kitchell, 1989).

Unfortunately, most restoration/rehabilitation schemes fail to or inadequately consider their actions within the context of horizontal lake management. It is usually assumed that littoral zones and broader fringing wetlands, like the pelagic, will return to their former extent, structure and function through a management plan. As demonstrated for Lakes Koronia and Chimaditida in Greece, failure to consider changes in physical and chemical aspects of the "memory" (sediments, soils) may invalidate such assumptions.

Wetlands must not be considered as monotypic habitats interacting with lakes in direct proportion to their aerial extent. Extensive wetlands surrounding lakes can be divided into three distinct zones: (1) an upland-wetland ecotone, (2) a wetland-pelagic ecotone and (3) an interior core area that rarely interacts directly with either upland or lake (Crisman et al., 2003a). As illustrated by Lake Chimaditida, there is likely to be a differential response of these zones to elevated water levels, especially after a prolonged period of lower water and sediment exposure. Proposals to harvest reeds from fringing wetlands to manage nutrient loading to Greek lakes (Nikolaidis et al., 1996), for example, may prove largely ineffective in lakes like Chimaditida if the wetland zone in question is so large and isolated that it rarely interacts with the pelagic zone of the lake.

The relative importance of vertical versus horizontal lake management aspects in overall lake rehabilitation schemes is governed by a number of factors including basin morphometry in conjunction with current and projected lake depth and structural/functional aspects of macrophytes versus phytoplankton communities. Above all, it is important to develop sound management goals for the rehabilitated lake that include both terrestrial and aquatic aspects. A predominately vertical lake management approach is probably valid for systems such as Lake Koronia without a history of significant submersed or emergent macrophytes. For those lakes embedded within significant wetlands like Lake Chimaditida; however, failure to consider horizontal lake management as a significant component of the overall system rehabilitation will likely diminish its successful outcome.

Nations throughout the Balkans, Near East, and Middle East are facing the unpleasant reality that there is likely not to be sufficient available fresh water resources to return lakes to previous water levels. They must establish clear objectives for both pelagic and littoral/wetland lake areas and strive to achieve these by determining how little water is needed both to manage the structure and function of regional lakes and to determine critical times annually when it must be present. Finally, definitions of wetlands currently used by Ramsar and aquatic scientists based primarily on structural aspects of ecosystems need to be modified to recognize the overriding importance of aerially differentiated functional aspects within vegetated communities as well as fundamental differences between vegetated and open-water habitats.

References

Anthemidis, A., Zachariadis, G., Stratis, I., Voulgaropoulos, A., Vasilikiotis, G., 1997. Analytical determination of heavy metals in Lake Koronia sediments. Proc. Hell. Symp. Oceanogr. Fish 2, 313–316.

Bayar, A., Soyupak, S., Altinbilek, D., Mukhallalati, L., Kutoglu, Y., Goekcay, C.F., 1997. Use of modelling for development of management strategies in control of eutrophication for systems of Mogan and Eymir. Fresenius Environ. Bull. 6, 115–120.

Beklioglu, M., Moss, B., 1996. Existence of a macrophyte-dominated clear water state over a very wide range of nutrient concentrations in a small shallow lake. Hydrobiologia 337, 93–106.

Brenner, M., Keenan, L.W., Miller, S.J., Schelske, C.L., 1999a. Spatial and temporal patterns of sediment and nutrient accumulation in shallow lakes of the Upper St. Johns River Basin, Florida. Wetlands Ecol. Manag. 6, 221–240.

Brenner, M., Whitmore, T.J., Curtis, J.H., Hodell, D.A., Schelske, C.L., 1999b. Stable isotope (^{13}C and ^{15}N) signatures of sedimented organic matter as indicators of historic lake trophic state. J. Paleolimnol. 22, 205–221.

Calzada-Bujak, I., Serrano, L., Toja, J., Crisman, T.L., 2001. Phosphorus dynamics in a Mediterranean temporary pond, Donana National Park, Spain. Verh. Int. Verein. Limnol. 27, 3986–3991.

Carpenter, S.R., Kitchell, J.F. (Eds.), 1989. The Trophic Cascade in Lakes. Cambridge University Press, Cambridge, p. 385.

Chapman, C.A., Chapman, L.J., 2003. Deforestation in tropical Africa: impacts on aquatic ecosystems. In: Crisman, T.L., Chapman, L.J., Chapman, C.A., Kaufman, L.S. (Eds.), Conservation, Ecology, and Management of African Freshwaters. University Press of Florida, Gainesville, FL, pp. 229–246.

Chapman, L.J., Chapman, C.A., Nordlie, F.G., Rosenberger, A.E., 2002. Physiological refugia: swamps, hypoxia tolerance and maintenance of fish biodiversity in the Lake Victoria region. Comp. Biochem. Physiol. 133(A), 421–437.

Cooke, G.D., Welch, E.B., Peterson, S.A., Newroth, P.R., 1993. Restoration and Management of Lakes and Reservoirs. Lewis Publishers, Boca Raton, 548 pp.

Crisman, T.L., 1986. Eutrophication control with an emphasis on macrophytes and algae. In: Polunin, N. (Ed.), Ecosystem Theory and Application. Wiley Press, pp. 200–239.

Crisman, T.L., Beaver, J.R., 1990. A latitudinal assessment of distribution patterns in chaoborid abundance for eastern North American lakes. Verh. Int. Verein. Limnol. 24, 547–553.

Crisman, T.L., Chapman, L.J., Chapman, C.A., 2003a. Incorporating wetlands and their ecotones in the conservation and management of freshwater ecosystems of Africa. In: Crisman, T.L., Chapman, L.J., Chapman, C.A., Kaufman, L.S. (Eds.), Conservation, Ecology, and Management of African Freshwaters. University Press of Florida, Gainesville, FL, pp. 210–228.

Crisman, T.L., Chapman, L.J., Chapman, C.A., Kaufman, L.S. (Eds.), 2003b. Conservation, Ecology and Management of African Fresh Waters. University Press of Florida, Gainesville, FL, 514 pp.

Deppe, T., Benndorf, J., 2002. Phosphorus reduction in a shallow hypereutrophic reservoir by in-lake dosage of ferrous iron. Water Res. 36, 4525–4534.

Dittrich, M., Koschel, R., 2002. Interactions between calcite precipitation (natural and artificial) and phosphorus cycle in the hardwater lake. Hydrobiologia 469, 49–57.

Economidis, P.S., Sinis, A.I., Stamou, G.P., 1988. Spectral analysis of exploited fish populations in Lake Koronia (Macedonia, Greece) during the years 1947–1983. Cybium 12, 151–159.

Fotis, G., Conides, A., Koussouris, T., Diapoulis, A., Gritzalis, K., 1992. Fishery potential of lakes in Macedonia, North Greece. Fresenius Environ. Bull. 1, 523–528.

Frankiewicz, P., Dabrowski, K., Zalewski, M., 1996. Mechanisms of establishing bimodality in a size distribution of age-0 pikeperch, Stizostedion lucoperca, in Sulejow Reservoir, central Poland. Annales Zoologici Fennici. 33 (3–4), 321–327.

Frazier, S., 1996. Directory of Wetlands of International Importance—An Update. Ramsar Convenstion Bureau, Gland, Switzerland, 236 pp.

Gophen, M., 2000. Nutrient and plant dynamics in Lake Agmon wetlands (Hula Valley, Israel): a review with emphasis on Typha domingensis (1994–1999). Hydrobiologia 400, 1–12.

Green, A.J., Fox, A.D., Hilton, G., Hughes, B., Yarar, M., Salathe, T., 1996. Threats to Burdur Lake ecosystem, Turkey, and its waterbirds, particularly the white-headed duck *Oxyura leucocephala*. Biol. Conserv. 76, 241–252.

Hansen, J., Reitzel, K., Jensen, H.S., Andersen, F.O., 2003. Effects of aluminum, iron, oxygen and nitrate additions on phosphorus release from the sediment of a Danish Softwater Lake. Hydrobiologia 492, 139–149.

Hellenic Ministry of Environment, Physical Planning and Public Works: Environmental Planning Division (HMEPPPW), 1996. Program of Management of the Protected Area of Lakes Koronia, Volvi and Their Surrounding Areas. Prefectures of Thessalonikiand Chalkidiki. Part A. Athens, Greece (in Greek).

Hollis, G.E., Stevenson, A.C., 1997. The physical basis of the Lake Mikri Prespa systems: geology, climate and water quality. Hydrobiologia 351, 1–19.

Hovhanissian, R., Gabrielyan, B., 2000. Ecological problems associated with biological resource use of Lake Sevan. Armenia. Ecol. Eng. 16, 175–180.

Junk, W.J., Bayley, P.B., Sparks, R.E., 1989. The flood pulse concept in river-floodplain systems. In: Dodge, D.P. (Ed.), Proceedings of the International Large River Symposium, Department of Fisheries and Oceans, Ottawa. Can. Spec. Publ. Fish. Aquat. Sci. vol. 106, 110–127.

Koussouris, T.S., Bertahas, I.T., Diapoulis, A.C., 1992. Background trophic state of Greek Lakes. Fresenius Environ. Bull. 1, 96–101.

Loeffler, H., Schiller, E., Kusel, E., Kraill, H., 1998. Lake Prespa, a European natural monument, endangered by irrigation and eutrophication? Hydrobiologia 384, 69–74.

Mıtrakı, C., Crisman, T.L., Zalidis, G., 2004. Lake Koronia: shift from autotrophy to heterotrophy with cultural eutrophication and progressive water-level reduction. Limnologica 34, 110–116.

Mitsch, W.J., Day Jr., J.W., Wendell Gilliam, J., Groffman, P.M., Hey, D.L., Randall, G.W., Wang, N., 2001. Reducing nitrogen loading to the Gulf of Mexico from the Mississippi River Basin: strategies to counter a persistent ecological problem. BioScience 51, 373–388.

Newbold, J.D., Elwood, J.W., O'Neil, R.V., Van, W., Winkle, 1981. Measuring nutrient spiraling in streams. Can. J. Fish. Aquat. Sci. 38, 860–863.

Nikolaidis, N.P., Koussouris, T., Murray, T., 1996. Seasonal variation of nutrients and heavy metals in *Phragmites australis* of Lake Trichonis, Greece. Lake Reserv. Manag. 12, 364–370.

Psenner, R., 1994. environmental impacts on fresh-waters—acidification as a global problem. Sci. Total Environ. 143, 53–61.

Ramsar Convention Secretariat. 2004. The Ramsar Convention Manual: a Guide to the Convention on Wetlands (Ramsar, Iran, 1971), third ed., Gland, Switzerland.

Reddy, K.R., Delaune, R.D., Debusk, W.F., Koch, M.S., 1993. Long-term nutrient accumulation rates in the Everglades. Soil Sci. Soc. Am. J. 57 (4), 1147–1155.

Reitzel, K., Hansen, J., Jensen, H.S., Andersen, F.O., Hansen, K.S., 2003. Testing aluminum addition as a tool for lake restoration in shallow, eutrophic Lake Sonderby Denmark. Hydrobiologia 506, 781–787.

Rosenberger, A.E., Chapman, L.J., 1999. Hypoxic wetland tributaries as faunal refugia from an introduced predator. Ecol. Freshwater Fish 8, 22–34.

Salonen, V.P., Varjo, E., Rantala, P., 2001. Gypsum treatment in managing the internal phosphorus load from sapropelic sediments; experiments on Lake Laikkalammi. Finland. Boreal Environ. Res. 6, 119–129.

Scheffer, M., 1998. Ecology of Shallow Lakes. Chapman and Hall, London, 375 pp.

Shih, S.F., Glaz, B., Barnes, R.E.B., 1998. Subsidence of organic soils in the Everglades Agricultural Area during the past 19 years. In: Soil and Crop Science Society of Florida Proceedings, vol. 57, pp. 20–29.

Soltero, R.A., Sexton, L.M., Ashley, K.I., McKee, K.O., 1994. Partial and full lift hypolimnetic aeration of Medical Lake, WA to improve water quality. Water Res. 29, 2297–2308.

Trepagnier, C.M., Kogas, M.A., Turner, R.E., 1995. Evaluation of wetland gain and loss of abandoned, agricultural impoundments in South Louisiana, 1978–1988. Restor. Ecol. 3(4):299–303.

United Nations, 2003. Report of the World Summit on Sustainable Development, Johannesburg, South Africa, 26 August–4 September 2002. A/CONF.199/20*. United Nations, New York, 173 pp.

Webster, J.R., Patten, B.C., 1979. Effects of watershed perturbation on stream calcium, potassium and calcium dynamics. Ecol. Monogr. 49, 51–72.

Wetzel, R.G., 2001. Limnology: Lake and River Ecosystems. Academic Press, San Diego, 1006 pp.

Wiley, D., 1997. Study of Hillsborough County lake enhancement complete. Land Water 41, 23–25.

Zalidis, C.G., Mantzavelas, A.L. (Eds.), 1994. Inventory of Greek Wetlands as Natural Resources. Greek Biotope/Wetland Centre (EKBY). Thermi, Greece, 448 pp.

Available online at www.sciencedirect.com

ECOLOGICAL ENGINEERING

ELSEVIER   Ecological Engineering 24 (2005) 391–401

www.elsevier.com/locate/ecoleng

Pantanal: a large South American wetland at a crossroads

Wolfgang J. Junk [a,*], Catia Nunes de Cunha [b]

[a] Max-Planck-Institute for Limnology, Working Group Tropical Ecology, PB 165, 24306 Plön, Germany
[b] Universidade Federal de Mato Grosso (UFMT), Depto. de Botânica e Ecologia, Instituto de Biociências,
Av. Fernando Correa, 78060-900 Cuiabá-MT, Brasil

Received 22 April 2004; received in revised form 8 November 2004; accepted 10 November 2004

Abstract

The Pantanal, a large and still rather pristine wetland in the center of the South American continent, is becoming increasingly threatened by large development programs. Agroindustries and reservoirs for hydroelectric power generation in the catchment area modify discharge pattern and sediment load of the tributaries, plans for canalization of the Paraguay River (hidrovia) are putting in risk the natural flood regime of large areas inside the Pantanal, and attract industries with high potential for environmental pollution, economic pressure on the traditional cattle ranchers accelerates the transformation of natural vegetation into pasture, etc. These activities negatively affect habitat and species diversity and scenic beauty but also the hydrological buffer capacity of the Pantanal. The article summarizes the ecological conditions of the Pantanal, discusses commercial and non-commercial values of the area, describes constraints for the development of intensive agriculture and cattle ranching, and discusses development alternatives. Considering the low density of human population inside the Pantanal, it can be concluded that development pressure on the Pantanal arises mostly from pressure groups outside the area that will also mostly benefit from the economic return of the development projects. Low density of human population would still allow the application of economically viable and environmentally friendly development alternatives that maintain and sustainably manage one of the largest wetlands in the world.
© 2005 Published by Elsevier B.V.

Keywords: Pantanal; Floodplain; Sustainable management

1. Introduction

The Pantanal is a large wetland of about 160,000 km^2 in the center of the South American continent. Its isolation from major consumption centers and the difficult access to the ranches inside the vast floodplain hindered economic development since colonization by Europeans at the beginning of the 18th century. The low human population density and the extensive cattle ranching had little impact on the environment. Therefore, the Pantanal today is still in a rather pristine condition.

During the last decades, changing economic and political requirements increased the pressure on the Pantanal and its catchment area. In recent years, the governments of Brazil, Bolivia, and Paraguay, in which the Pantanal lies, have made major efforts to involve

---

* Corresponding author.
  E-mail address: wjj@mpil-ploen.mpg.de (W.J. Junk).

0925-8574/$ – see front matter © 2005 Published by Elsevier B.V.
doi:10.1016/j.ecoleng.2004.11.012

the Pantanal in national economic development. This is particularly evident for Brazil, which holds approximately 85% of the area (Alho et al., 1988). Developmental projects profess to improve living conditions and stimulate economic growth, with far-reaching ecological and socio-economic consequences; however, their consequences have not been analyzed in detail.

On the other hand, the local population, scientists, governmental and non-governmental agencies, and politicians are increasingly trying to find ways to protect the Pantanal and to maintain its unique natural resources. The Pantanal is considered "globally outstanding" (rank 1 of 4) in terms of biological distinctiveness and "vulnerable" (rank 3 of 5) in terms of conservation, and has "highest priority" (rank 1 of 4) in regional priorities for conservation action according to a conservation assessment of the WWF and the Biodiversity Support Program (Olson et al., 1998).

Discussions on the future of the Pantanal already started some years ago, but such discussions are hindered by insufficient databases on economic, socioeconomic, hydrological, and ecological factors, as illustrated by the discussion of the *hidrovia* (Huszar et al., 1999; Gottgens et al., 2001). Increasing economic and political pressure requires fundamental decisions to be made in the near future, and the Pantanal, indeed, is now at a crossroads.

The aim of this article is to summarize the knowledge on major structures and functions of the Pantanal, to describe the role of the flood pulse for the ecosystem, and to delineate some important ecosystem functions for the well being of the human population. The major human impacts on the ecosystem will also be described and various developmental schemes will be discussed.

1.1. Ecological outlines

The Pantanal is situated in the Alto Paraguay Depression, which extends between the young uplifting Andes in the west and the old crystalline Central Brazilian Shield in the east (Fig. 1). The main phase of the subsidence that resulted in the wetland depression very likely occurred during the upper Pliocene to lower Pleistocene Epochs about 2.5 million years ago (Adámoli, 1981; Barros, 1982; Del'Arco et al., 1982; Alvarenga et al., 1984; Godoi Filho, 1986). Positioned 15–20° south of the Equator, the area is situated in a circumglobal belt of climate instability and was subject to severe climatic changes during the Quaternary Period. Alternating dry and wet periods led to different patterns of discharge and sediment load of the Paraguay River and its tributaries, which resulted in a mosaic of different geomorphologic formations that are covered today by various types of vegetation (Short and Blair, 1986; Jimenez-Rueda et al., 1998). During the late Pleistocene and Holocene, the Pantanal passed through several changes between wet and dry episodes as follows: 40,000–8000 BP cold and dry, 8000–3500 BP warm and wet, 3500–1500 warm and dry and 1500-Present warm and wet (Ab'Saber, 1988; Iriondo and Garcia, 1993; Stevaux, 2000).

During paleo-climatic dry periods, extinction rates of wetland organisms were high. Re-immigration from the lower Paraguay River, the surrounding Cerrado, the Amazon basin, and the Chaco occurred (Fig. 1). Mobile species, such as aquatic birds and insects, were favored. The time span since the last dry period was obviously not long enough for the development of endemic species. Furthermore, flood pulse induced migration, and passive transport of organisms hindered speciation by spatial segregation of populations.

Today, the Pantanal is a wetland subject to a predictable monomodal flood pulse (Fig. 2). This pulse is the driving force in the Pantanal landscape (Junk and Da Silva, 1999; Junk, 2000). The considerable annual and multi-annual variability affects the biota with different intensities and on different time scales (Fig. 3, Nunes da Cunha and Junk, 2004). The vast plain stores water during the rainy season and delivers it slowly to the lower sections of the Paraguay River, thereby buffering its flood amplitude. During the passage through the Pantanal, about 90% of the water returns to the atmosphere, contributing considerably to the regional water and heat balance (Ponce, 1995). Any long-term change of the pulse will result in fundamental ecological changes in the affected areas and also influence the living conditions of the local human population. Wildfires and human-induced fires represent important additional stresses to the Pantanal that also affect fauna and flora with different intensities and on different time scales. The long-term impact of fire on the distribution and abundance of the various organisms is not yet fully understood.

A large diversity of habitats leads to a broad species diversity. The Pantanal harbors only very few endemic species, but large populations of charismatic South

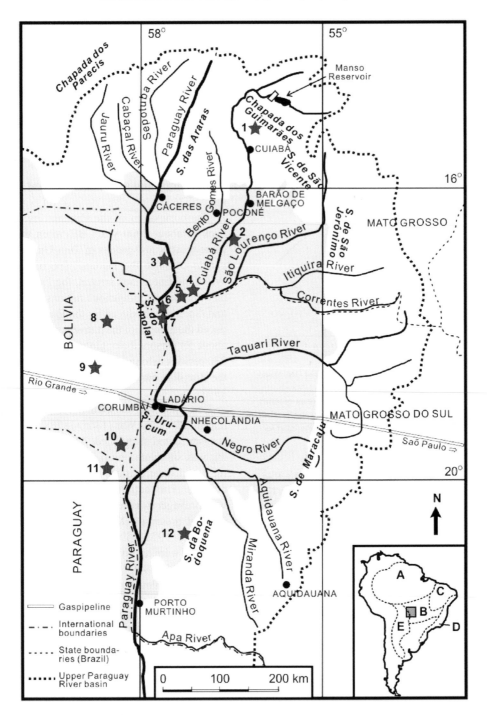

Fig. 1. Map of the Pantanal and its catchment area and position of protected areas. 1 = PN Chapada dos Guimarães, 2 = RPPN-SESC Pantanal, 3 = EE Taiamã, 4 = RPPN Dorochê, 5 = PN do Pantanal, 6 = RPPN Acurizal, 7 = RPPN Penha, 8 = ANMI San Matias, 9 = Reserva Municipal del Valle de Tucavaca, 10 = PN-ANMI Otuquis, 11 = PN Rio Negro, 12 = PN Serra da Bodoquena. The small map indicates the position of the Pantanal in South America and the surrounding biomes. A = Amazon forest, B = Cerrado, C = Caatinga, D = Atlantic forest, E = Chaco. Further explications in text.

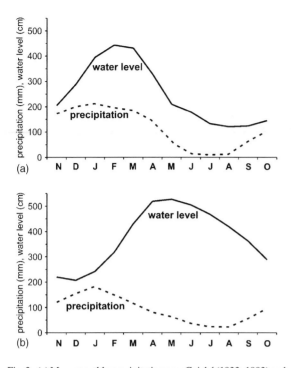

(a)

(b)

Fig. 2. (a) Mean monthly precipitation near Cuiabá (1933–1993) and mean water level of the Cuiabá River at Cuiabá (1971–1988), northern Pantanal (according to Zeilhofer, 1996); and (b) mean monthly precipitation near Corumbá (1912–1971) and mean water level of the Paraguay River at Ladário (1979–1987), southern Pantanal (according to Hamilton et al., 1999).

American species that are threatened outside the Pantanal by extinction are found there (Da Silva, 2000; Junk et al., 2005a). Diversity of landscape units gives the Pantanal a high aesthetic value, i.e., as parkland

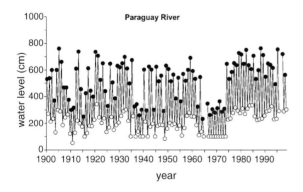

Fig. 3. Annual water level fluctuations of the Paraguay River at Ladário from 1900 to 2000 (data according to DNAEE): (●) maxima; (○) minima.

landscape. The human population density in the Pantanal is low and concentrated in small urban areas at the non-flooded borders of the Pantanal, along the major rivers, and in the ranches scattered over the floodplain. Most soils of the Pantanal are acidic and of low fertility (Amaral Filho, 1986). Their agricultural potential is low. Furthermore, the correlation of rainfall with flooding hinders the planting of dry-land crops.

2. Economic and social development

The occupation of the Pantanal by humans dates back to about 5000 years BP, when the climate became moister and groups of Tupi-Guarani Indians began to colonize the Pantanal (Peixoto et al., 1999). When the Europeans arrived, the Pantanal was occupied by various indigenous nations. Wars, slave rides and diseases introduced by the Europeans quickly reduced the number of the native population. Today, only about 50 persons belonging to the Guató Nation and 270 persons belonging to the Bororo Nation live inside the Brazilian part of the Pantanal (Da Silva and Silva, 1995).

Cattle ranching started in the mid-18th century. During the Paraguay War (1865–1870), the ranches were devastated and abandoned. After the war, cattle ranching was stimulated by the export of salted and dried meat and meat extracts to national and international markets. This activity declined after World War II, when refrigeration techniques decreased the demand of meat preserves (Mazza et al., 1994; Pasca, 1994; Remppis, 1995). Environmental impacts of cattle ranching were relatively low.

In the mid-1970s, the Brazilian government started several large development programs that affected the Pantanal, such as the Intermunicipal Consortium for the Development of the Pantanal (CIDEPAN), the Program for the Development of the Pantanal (PRODE-PAN), the Program for the Development of the Cerrados (POLOCENTRO), the National Alcohol Program (PROÁLCOOL), the Development Program of the Grande Dourados (PRODEGRAN), the Study of the Integrated Development of the Upper Paraguay Basin (EDIBAP), the Integrated Program of the Development of the North-East of Brazil (POLONOROESTE), the Program of the Agro-Environmental Development of the State of Mato Grosso (PRODEAGRO),

and the National Environmental Program (PNMA), with its sub-program Conservation Plan of the Upper Paraguay Basin (PCBAP) (Alho et al., 1988; Junk et al., 2005b).

The aims of the programs were to intensify the utilization of the natural resources of the Pantanal and its catchment area and to integrate the region into the national development scheme, for instance, by the construction of roads and lines for electric energy transmission. Indeed, they stimulated the agroindustrial development of the region, but also brought about serious negative ecological side effects for the savanna vegetation in the catchment area (Cerrado) and the Pantanal. Competition with cattle ranching on artificial pastures in the Cerrado place economic pressure on traditional ranches to increase beef production that affects the ecosystem, for instance, by overgrazing, deforestation for the increase of pasture areas, and plantation of artificial pastures. The sediment load of the tributaries, such as the Taquarí River, is rising because of increased erosion caused by large agroindustrial projects in the surrounding upland (*chapadas*). Since the 1980s, gold mining in the lowlands near the city of Poconé releases mercury in the environment, but superficial gold deposits are now exhausted and mining activities have sharply declined during the last years.

Actually there are nine hydroelectric power plants with a total capacity of 323 MW operating in the Pantanal catchment area but only the one on the Manso River, tributary to the Cuiabá River, is of large size (area 387 km^2, capacity 220 MW) (Fig. 1). Changes in hydrology caused by the large Manso River reservoir, begin to affect flora, fauna and also fishermen and cattle ranchers along the Cuiabá River inside the Pantanal. Reservoir number may rise in future to 31 with a total capacity of 1064 MW, three of them of large size at Rio Correntes (176 MW), Rio Itiquira (156 MW) and Rio Jauru (110 MW). An environmental impact analysis about the cumulative effect of the projected reservoirs on the Pantanal shows that the construction of large reservoirs should be avoided, because they strongly modify the hydrological regime of the affected rivers (Girard, 2002).

The projected Bolivia–Brazil gas pipeline from Rio Grande in Bolivia to Sao Paolo, Rio, de Janeiro, Campos, Belo Horizonte, Curitiba, Florianopolis and Porto Alegre will pass the Pantanal from Corumbá to Campo Grande (Fig. 1). The gas will be used for the generation of thermoelectric energy and there are plans for a large gas-chemical complex in Corumbá.

A major threat is the economic pressure, placed by agro-businesses and mining industry outside the Pantanal, to canalize the Paraguay River for inexpensive commercial navigation of soybeans and minerals to the Atlantic Ocean (*hidrovia*); this would lead to large-scale, irreversible wetland degradation and seriously affect the living conditions of the local human population (Ponce, 1995; Hamilton, 1999). The multiple interactions between man and environment in the Pantanal are shown in Fig. 4.

In Brazil, increasing concern about the future of the Pantanal has led to a variety of activities of the Federal Universities of Mato Grosso (UFMT) and Mato Grosso do Sul (UFMS), the State Universities of Mato Grosso (UNEMAT) and Mato Grosso do Sul (UEMS), Brazilian Ministry for the Environment (IBAMA), State Agency for the Environment (SEMA), Agricultural Research Unit in Corumbá (UEPAE) under the leadership of Brazilian Agricultural Research Agency (EMBRAPA), Institute of the Defense of Agriculture and Animal Ranching (INDEA), States Secretariat of the Environment and Sustainable Development of Mato Grosso do Sul (SEMADES) and others. Furthermore, several national and international non-governmental agencies are working in the area. In 1988, the Pantanal was declared by the Brazilian constitution as a National Heritage. In 1993, UNESCO declared it as a Ramsar Site, and in 2000 as a World Biosphere Reserve and granted it the World Heritage Certificate. In 2002, the Pantanal Regional Environmental Program, part of the United Nations University (UNU/PREP), was founded at the Federal University of Mato Grosso. UNU/PREP leads a consortium of local universities and professes to establish a network of national and foreign institutions interested in the sustainable management and protection of the Pantanal.

There are two National Parks and one Ecological Station under governmental administration and several private protected sites inside the Brazilian part of the Pantanal. Two major private protected sites are administrated by the NGO ECOTROPICA and the Social Service of Commerce (SESC). The total protected area inside the Pantanal consists of 360,000 ha, corresponding to about 2.6% of the Brazilian part of the Pantanal (http://www.ibama.gov.br/). The National Park Rio Negro in Paraguay covers 123,786 ha. In the Bo-

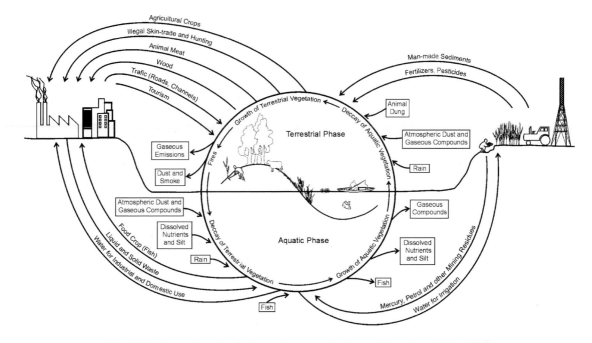

Fig. 4. Nutrient cycles and major human impacts on the Pantanal (according to Junk, 2002).

livian Pantanal, there are three protected areas: the Natural Area of Integrated Management San Matías (ANMI San Matías, 2,918,500 ha), the National Park and Area of Integrated Management Otuquis (PN-ANMI Otuquis, 1,005,950 ha), and the Municipality Reserve of Tucavaca (262,305 ha). These areas include wetlands but also uplands in different proportions (http://www.fobomade.org.bo/pantanal_bolivia/conociendo.php) (Fig. 1).

A research unit (Unidade de Execução de Pesquisa de Âmbito Estadual de Corumbá, UEPAE) under the leadership of the Brazilian Agricultural Research Agency (EMBRAPA) at Corumbá provides technical assistance for agriculture and cattle ranching inside the Pantanal. Researchers of the UFMT have been studying the Pantanal since the 1970s and have cooperated with the Max-Planck-Institute for Limnology in Plön, Germany, since 1991 on ecological research of the Pantanal, capacity building, and development of concepts for sustainable development. In 2002, the Pantanal Ecology Research Group (NEPA) was founded at the UFMT, and the Ecological Research Program of Long Duration (PELD) was established at the SESC Reserve.

3. Globalization and development alternatives

Today, many countries of the tropical and subtropical belt face similar problems to those occurring in the Pantanal and try to combine environmental protection with economic needs, social justice, and regional development requirements in wetland management. The complexity of the problem is illustrated by the rather small number of success stories and the large number of projects with heavy negative ecological, economic, and social side effects (Junk, 2002). In most African and Asian countries, the problem is aggravated by high population growth rates in and around the wetlands, the increased water requirement of urban and industrial centers, unsuitable agricultural practices, and inefficient energy use.

Countries bordering the Pantanal have the privilege of determining the future of one of the most famous wetlands under a low political pressure level: (1) the ecosystem is still in good health, (2) the population density in and around the Pantanal is very low, (3) in contrast to most African and some Asian countries that suffer from dictatorship, the democratic governments of the countries lying partly in the Pantanal can rely on

a well-developed network of governmental and non-governmental organizations that allow an efficient exchange of information between politicians, planners, scientists, the local population, and other stakeholders. This should give planners and decision makers the freedom to collect as much information as necessary, to analyze carefully the pros and cons of the various developmental alternatives, and to establish a long-term developmental policy for the sustainable use of the Pantanal and its natural resources by including the local population in the decision-making process.

There is no doubt that a sustainable development of the Pantanal should be achieved. But what is sustainable development? The term "sustainable" was defined by the Consultative Group on International Agricultural Research (CGIAR) as the "successful management of resources for agriculture to satisfy changing human needs while maintaining or enhancing the quality of the environment and conserving natural resources" (Food and Agriculture Organization (FAO), 1989). This definition can be also used for wetlands, but without restriction to agriculture because wetlands can be managed for multiple purposes according to regional needs. The requirement that "sustainable development should be ecologically sound, economically viable, socially just, culturally appropriate and based on a holistic scientific approach" reflects the concern of many environmentalists, development agencies, and politicians not to separate society and environment, and economy and ethics (Becker, 1997). However, these definitions are too vague to be put into practice.

The real-world situation of the Pantanal and the necessity to develop feasible and viable activities for a "sustainable" development of the Pantanal require a detailed analysis of the "natural capital" represented by the natural resources of the Pantanal, an estimate of their development potential, a socio-economic impact study, and an ecological risk assessment for these activities. The "natural capital" considered in this analysis should not be based only on the economic value of single resources, as for instance, fish, wood, and beef production, but also on the value of ecological services, such as water and sediment retention, water purification, stabilization of regional climate, maintenance of biodiversity, and quality of life for the local population (Daily and Ehrlich, 1996; Daly, 1991; Goodland, 1991). The value of these services is often seriously underestimated because they do not provide an imme-

diate cash return and therefore suffer the "tragedy of the commons" (Hardin, 1968): everyone wants to have clean air, clear water, beautiful landscape, and wildlife, but no one wants to pay for and take care of them because they are considered common goods.

The importance of wetlands is underlined by the following average values for ecosystem goods, services, biodiversity, and cultural considerations that have been estimated for various ecosystems: US$ 8498 ha^{-1} yr^{-1} for rivers and wetlands, US$ 969 ha^{-1} yr^{-1} for forests, and US$ 232 ha^{-1} yr^{-1} for grasslands (Constanza et al., 1997). In the past, highly industrialized countries such as the USA and European countries have invested large amounts of money to "valorize" their wetlands by using them for agriculture and the construction of infrastructure, industries, and housing. Negative economic and ecologic side effects now force the USA government to spend many billions of US$ to recover parts of the Everglades and the Mississippi floodplain to benefit, in the long term, economically from the associated wetland values that are now highly appreciated (Abramovitz, 1996; TNC, 1996). Similar efforts have been taken by France, Germany, and the Netherlands to recover parts of the Rhine River floodplain (Nienhuis et al., 1998). These examples show that (1) economic benefits of wetland destruction are often overrun by the costs of negative side effects, (2) the economic framework changes quickly and modifies cost–benefit analyses of development projects, often in favor of the values of an intact wetland, (3) only parts of the former wetland area can be recovered to near-natural conditions with very high costs, and (4) maintenance of wetlands is always much less expensive and more effective than rehabilitation after degradation.

Several international conventions have set up a framework for the conservation of the Pantanal as a globally important wetland, e.g., the Ramsar Convention, the Convention on Biological Diversity, the Convention on Climate Change, the Convention on Migratory Species, and the World Heritage Convention. These conventions stress the importance of non-commercial values of ecosystems and allow the inclusion of the Pantanal in international networks with similar goals that facilitate exchange of information and access to international funding and political support.

The political decisions made in the near future will determine in which direction the Pantanal will develop.

One option is cautious development avoiding drastic changes, respecting the dynamic behavior of the ecosystem (Jansson and Jansson, 1994), and prioritizing sustainability when using the natural resources of the Pantanal. This option would consider the importance of the Pantanal as a hydrological buffer system for downriver areas, a regional climate buffer, a valuable area for water retention and purification, and a center of maintenance of biodiversity. Cattle ranching is a weak economic basis for the ranches, but the scenic beauty of the landscape and its diverse flora and fauna in combination with traditional extensive ranching is very attractive for different forms of ecotourism and sport fishing. Large-scale marketing campaigns for local "green" products such as for beef with a recognized label of "free roaming sustainably managed Pantanal cattle" could increase the profitability of the ranches. The viability of such concepts is shown in Bonito, a city at the southern edge of the Pantanal. Several small rivers with transparent water, a diverse fish fauna and a large variety of aquatic plants attract many tourists.

The other option is the well-known approach of short-term, profit-oriented, non-sustainable stimulation of agriculture and intensified cattle ranching by large-scale road construction and flood control, and stimulation of industrial production by the construction of large hydroelectric power plants, canalization of the Paraguay River for large ship traffic, and other measures to improve infrastructure. This option would strongly affect and in part destroy one of the largest and best-preserved wetland systems of the world, and the economical return is questionable. Globalization of the markets will further increase the economic pressure and accelerate development projects in the Pantanal. Of special concern is the "tyranny of small decisions" because small steps taken are not very apparent to the public, but their sum can have a considerable negative effect. These seemingly small, but important steps should be carefully monitored and controlled by the environmental organizations.

For example, the disastrous consequences of the construction of the *hidrovia* on the Pantanal have been discussed by various authors (Ponce, 1995; Huszar et al., 1999; Gottgens et al., 2001). The construction requires meanders of the Paraguay River to be cut off, the river channel to be deepened, buildings to be constructed along its shores, and, most strikingly, rocky outcrops in the river channel to be removed, these serv-

ing as a sequence of natural impoundments and thus regulating the extent of the inundation (Ponce, 1995). Changes in hydrology would be irreversible.

Hamilton (1999) has conservatively estimated that a lowering of the river channel depth by 10 or 25 cm could reduce the flooded area by 11.7 or 31.4%, respectively. Willink et al. (2000) concluded that with the elimination of the northern Pantanal wetlands e.g., by the construction of the *hidrovia*, areas with high fish species diversity containing species of high economic value would be destroyed and 40–60% of the species could be eliminated. However, despite a judgment to stop any activity until a final decision about the *hidrovia* has been made, large vessels are used on the Upper Paraguay River. They recurrently become stranded, damage the ecosystem and increase the pressure on the population to vote in favor of channel deepening and harbor construction (Wantzen et al., 1999).

Another example is the settlement of landless people from the "Movimento Sem Terra" (MST, Movement of Landless People) in the Pantanal. Periodical flooding does not allow the application of traditional land-use techniques and makes the Pantanal an unsuitable place for subsistence farming. Settlers of the "Movement of Landless People" in the area cannot rely on traditional agriculture, and ranchers report of poaching and bushfires caused by these people who are not accustomed to the conditions in the Pantanal. Daily and Ehrlich (1992, 1996) define "carrying capacity" as "the maximum population size of a given species that an area can support without reducing its ability to support the same species in the future". Of course, humans can increase the carrying capacity through adequate management methods; however, it must be admitted that the natural carrying capacity of the Pantanal for humans is low.

In a letter of 13 April of 2004 to Mrs. Marina Silva, Minister of the Environment, the NGO's Ecologia e Ação (Ecology and Action, ECOA) and Organização de Cultura, Cidadania e Ambiente (Organization of Culture, Citizenship and Environment, OCCA) expressed their deep concerns about plans to establish a metallurgical and gas-chemical complex in combination with a highly polluting thermoelectric power plant operating with natural gas from Bolivia in the city of Corumbá. Supporters of the project reinforce their argument citing economic development and the creation of jobs for the local population. But there are certainly much better options for job creation in environmen-

tally friendly activities in such a unique and sensitive wetland than the establishment of a highly polluting industry with heavy negative impact on environment, human health, tourism, fishery and other activities. Once the industrial complex is established, there is no return to former near-pristine conditions in the area. On the contrary: it will call for other polluting industries and accelerate destructive activities.

4. Summary and recommendations

In a world with intensifying intercontinental economic links and with the countries lying partly within the Pantanal having growing economies and an increasing human population with changing requirements for a standard of living, an appeal to keep the Pantanal untouched would be unrealistic. However, any developmental planning should carefully analyze whether the proposed developmental projects make sustainable use of the specific "natural capital" of the Pantanal, how the projects could suffer from ecosystem restrains, how the projects affect the living standard of the local population, and to what extent they could negatively affect or irreversibly destroy major wetland structures and functions.

The "natural capital" of the Pantanal consists of its unique landscape and its high biodiversity, including large populations of charismatic animals, some of which are in danger of extinction elsewhere. Some very important ecosystem services are periodic water storage and release, stabilization of the regional climate, water purification, and sediment trapping.

Various features of the Pantanal act as limiting factors for its agricultural development, such as the high ecosystem fragility, and heavy natural stress factors, e.g., low nutrient status, periodic flooding, periodic drought, and fire. These lead to variable, but generally low to moderate natural productivity (Junk and Da Silva, 1999; Junk, 2000). Natural plant and animal communities compensate for these constraints by high nutrient-use efficiency and strong fluctuations in population density. However, the surplus that can be sustainably collected by humans is rather small, as shown by the adjacent Amazon basin. In just a few centuries, a small number of European immigrants reduced the formerly large populations of river turtles, manatees, caimans, otters, and capybaras to very low

levels or even to near extinction (Junk and Da Silva, 1997).

There is no doubt that natural limiting factors to a certain extent can be compensated by adequate management methods (Goodland and Daly, 1996). However, conventional methods often fail in floodplains because of specific ecosystem behavior. For instance, improving productivity of soils by application of fertilizers during the low-water period is uneconomical because the periodic flooding lixiviates the nutrients. Furthermore, flooding coincides with the rainy season and hinders the planting of non-aquatic crops. Elimination of the impact of flooding by large-scale dike construction is not advisable because it would destroy the specific character of the ecosystem, negatively affect the adjacent floodplain areas, and diminish or threaten wetland ecosystem services. Early attempts to enclose areas of the Pantanal in the Camargo de Correia Island with dikes led to a prolonged moisture period within the dikes (rather than keeping the water out as expected) and a subsequent growth of woody weeds, which made a large area unusable for cattle ranching. There are many other examples of failed attempts to drive the wide amplitude of environmental variables of seasonal wetlands into the range where they could support agroindustrial systems. From these experiences, it would seem to be wise to develop a "floodplain-friendly" philosophy and to collect knowledge for a strategy that profits from the unique adaptations and life strategies of the organisms that have thrived successfully in these wetlands over the long course of evolution.

Sustainable development includes the maintenance of vital ecosystem structures and functions for the benefit of future generations. For instance, major modifications of the flooding regime, such as those predicted if the *hidrovia* is constructed, would lead to dramatic large-scale modifications of the Pantanal (Hamilton, 1999) and certainly do not fit within the concept of sustainable development. Economic benefits of the *hidrovia* are already today questionable and might vanish in the future because of a quickly changing global economy and the arising national transport alternatives; Pantanal degradation, however, is irreversible (Huszar et al., 1999).

Over the past two centuries, low-intensity cattle ranching has proven to be a sustainable management approach that maintains structures, functions, biodi-

versity, and the beauty of the landscape and is one of the very few examples of sustainable management of a tropical ecosystem introduced by Europeans. However, increasing economic pressure in the last decades requires changes in the traditional management concepts. Possibilities to increase the economic return of the ranches by amplifying pasture area through deforestation or through increasing animal density per unit area are limited. Viable alternatives are emerging, for instance, by the stimulation of a well organized ecotourism that benefits the local population, better exploitation of the little-used stocks of iliophagous fish species for human consumption, export of ornamental fishes, cultivation of game animals (e.g., caimans and capybaras), and a better marketing of local products under a "green" label that indicates environmentally sound production. The detailed analysis of developmental alternatives should lead finally to the formulation of an integrated master plan for the sustainable development of the Pantanal and its catchment area.

References

Abramovitz, J.N., 1996. Imperiled Waters, Impoverished Future: The Decline of Freshwater Ecosystems. World Watch Paper 128. Worldwatch-Institute, Washington, DC, USA, p. 80.

Ab'Saber, A.N., 1988. O Pantanal Mato-Grossense e a teoria dos refugios. Revista Barsileira de Geografia 50, 9–57.

Adámoli, J., 1981. O Pantanal e suas relações fitogeográficas com os cerrados. Discussão sobre o conceito "Complexo do Pantanal'. In: Congresso Nacional de Botânica, 32, Teresina. Sociedade Brasileira de Botânica, pp. 109–119.

Alho, C.J.R., Lacher, T.E., Conçales, H.C., 1988. Environmental degradation in the Pantanal ecosystem. Bioscience 38, 164–171.

Alvarenga, S.M., Brasil, A.E., Pinheiro, R., Kux, H.J.H., 1984. Estudo geomorfológico aplicado à Bacia do alto Rio Paraguai e Pantanais Matogrossenses. Boletim Técnico Projeto RADAM/BRASIL. Série Geomorfologia, Salvador 187, 89–183.

Amaral Filho, Z.P., 1986. Solos do Pantanal Mato-grossense. In: EMBRAPA/UEPAE/UFMS (Eds.), Anais do I Simpósio sobre Recursos Naturais e Sócio-econômicos do Pantanal. EMBRAPA/UEPAE/UFMS, Brasilia EMBRAPA-DDT, pp. 91–103.

Barros, A.M., 1982. Geologia. In: Ministerio das Minas e Energia, Secretaria Geral Projeto RADAM/BRASIL Folha SD21 Cuiabá (Levantamento de Recursos Naturais 26), Rio de Janeiro, Brasil.

Becker, B., 1997. Sustainability assessment: a review of values, concepts and methodological approaches. The World Bank. Issues Agric. 10, 1–63.

Constanza, R., d'Arge, R., de Groot, R., Farber, S., Grasso, M., Hannon, B., Limburg, K., Naeem, S., O'Neill, R.V., Paruelo, J., Raskin, R.G., Sutton, P., van den Belt, M., 1997. The value of the world's ecosystem services and natural capital. Nature 387, 253–260.

Daily, G., Ehrlich, P., 1992. Population, sustainability, and Earth's carrying capacity. BioScience 42, 761–771.

Daily, G., Ehrlich, P., 1996. Socioeconomic equity, sustainability, and Earth's carrying capacity. Ecol. Appl. 6, 991–1001.

Daly, H., 1991. Elements of environmental macroeconomics. In: Constanza, R. (Ed.), Ecological Economics: The Science and Management of Sustainability. Columbia University Press, New York, pp. 32–46.

Da Silva, C.J., Silva, J.A.F., 1995. No Ritmo das Àguas do Pantanal. Núcleo de Apio à Pesquisa sobre Populações Humanas e Àreas Ùmidas Brasileiras. USP, São Paulo, Brazil, p. 194.

Da Silva, C.J., 2000. Ecological basis for the management of the Pantanal—Upper Paraguay River Basin. In: Smits, A.J.M., Nienhuis, P.H., Leuven, R.S.E.W. (Eds.), New Approaches to River Management. Blackhuys Publishers, Leiden, The Netherlands, pp. 97–117.

Del'Arco, J.O., Silva, R.H., Tarapanoff, I., Freire, F.A., Pereira, L.G.M., Luz, D.S., Palmeirea, R.C.B., Tassinari, C.C.G., 1982. Geologoica. In: Levantamento de Recursos Naturais. Folha DE-21, Corumbá e parte da Faloha SE-20, vol. 20. Ministério da Minas Energia, Depto. Nac. da Prod. Min. Projeto RADAM/BRASIL, Rio de Janeiro, Brazil, pp. 25–160.

Food and Agriculture Organization (FAO), 1989. Sustainable agricultural production: implications for international Agricultural Research. Research and Development Paper No. 4. FAO, Rome, 131 pp.

Girard, P., 2002. Efeito cumulativo das barragens no Pantanal. Instituto Centro Vida, Campo Grande-MS, Rios Vivos, Brazil, 27 pp.

Godoi Filho, J.D., 1986. Aspectos geológicos do Pantanal Matogrossense e de sua área de influência. Anais do 1 Símposio sobre Recursos Naturais e Sócio-Economics do Pantanal. Documentos 5, EMBRAPA-Pantanal, Corumbá, 265 pp.

Goodland, R., 1991. The environment as capital. In: Constanza, R. (Ed.), Ecological Economics: The Science and Management of Sustainability. Columbia University Press, New York, pp. 168–175.

Goodland, R., Daly, H., 1996. Environmental sustainability: universal and non-negotiable. Ecol. Appl. 6, 1002–1017.

Gottgens, J.F., Perry, J.E., Fortney, R.H., Meyer, J.E., Benedict, M., Rood, B.E., 2001. The Paraguay-Parana Hidrovia: protecting the Pantanal with lessons from the past. BioScience 51, 301–308.

Hamilton, S.K., 1999. Potential effects of a major navigation project (Paraguay-Paraná-Hidrovia) on inundation in the Pantanal Floodplains. Regul. Rivers: Res. Manage. 15, 289–299.

Hamilton, S., Sippel, S., Calheiros, D., Melack, J., 1999. Chemical characteristics of Pantanal waters. In: EMBRAPA (Ed.), Anais do II Simpósio sobre Recursos Naturais e Sócio-econômicos do Pantanal, Corumbá. EMBRAPA, Corumbá, 1996, pp. 89–100.

Hardin, G., 1968. The tragedy of the commons. Science 162, 1243–1248.

Huszar, P., Petermann, P., Leite, A., Resende, E., Schnack, E., Schneider, E., Francesco, F., Rast, G., Schnack, J., Wasson, J., Gacia Lozano, L., Dantas, M., Obrdlik, P., Pedroni, R., 1999. Fact

or Fiction: A Review of the Hydrovia Paraguay-Paraná Official Studies. WWF, Toronto, 217 pp.

Iriondo, M.H., Garcia, N.O., 1993. Climatic variations in the Argentine plains during the last 18,000 years. Palaeogeogr. Palaeoclimatol. Palaeoecol. 101, 209–220.

Jansson, B.-O., Jansson, A.M., 1994. Ecosystem properties as a basis for sustainability. In: Jansson, A.M., Hammer, M., Folke, C., Constanza, R. (Eds.), Investing in Natural Capital: The Ecological Economics Approach to Sustainability. Island Press, Washington, DC, pp. 74–91.

Jimenez-Rueda, J.R., Pessotti, J.E.S., Mattos, J.T., 1998. Modelo para o estudo da dinâmica evolutiva dos aspectos fisiográficos dos pantanais. Pesq. Agropec. Bras. V. 33, 1763–1773.

Junk, W.J., 2000. The Amazon and the Pantanal: a critical Comparison and lessons for the future. In: Swarts, F.A. (Ed.), The Pantanal: Understanding and Preserving the World's largest Wetland. Paragon House, St. Paul, Minnesota, pp. 211–224.

Junk, W.J., 2002. Long-term environmental trends and the future of tropical wetlands. Environ. Conserv. 29, 414–435.

Junk, W.J., Da Silva, C.J., 1999: O "Conceito do pulso de inundação" e suas implicações para o Pantanal de Mato Grosso. In: EMBRAPA (Ed.), Anais do II Simpósio sobre Recursos Naturais e Sócioeconomicos do Pantanal. Manejo e Conservação. EMBRAPA, Corumbá, Brazil, pp. 17–28.

Junk, W.J., Da Silva, V.M.F., 1997. Mammals, reptiles and amphibians. In: Junk, W.J. (Ed.), The Central-Amazonian Floodplain: Ecology of a Pulsing System. Ecological Studies, 126. Springer Verlag, Berlin, pp. 409–418.

Junk, W.J., Nunes da Cunha, C., Wantzen, K.M., Petermann, P., Strüssmann, C., Marques, M.I., Adis, J., 2005a. Comparative biodiversity of large wetlands: the Pantanal of Mato Grosso, Brazil. Aquat. Sci., in press.

Junk, W.J., Da Silva, C.J., Wantzen, K.M., Nunes da Cunha, C., Nogueira, F., 2005b. The Pantanal of Mato Grosso: status of ecological research, actual use, and management for sustainable development. In: Maltby, E. (Ed.), The Wetlands Handbook. Blackwell Science, Oxford, UK, in press.

Mazza, M.C.M., Mazza, C.A.S., Sereno, J.R.B., Santos, S.A., Pellegrin, A.O., 1994. Etnobiologia e conservação do bovino pantaneiro. EMBRAPA, Corumbá, 61 pp.

Nienhuis, P.H., Leuven, R.S.E.W., Ragas, A.M.J., 1998. New Concepts for Sustainable Management of River Basins. Backhuys Publishers, Leiden, 374 pp.

Nunes da Cunha, C., Junk, W.J., 2004. Year-to-year changes in water level drive the invasion of *Vochysia divergens* in Pantanal grasslands. Appl. Veg. Sci. 7, 103–110.

Olson, D., Dinerstein, E., Canevari, P., Davidson, I., Castro, G., Morisset, V., Abell, R., Toledo, E., 1998. Freshwater biodiversity of Latin America and the Carribean: a conservation assessment. Biodiversity Support Program, Washington, DC, 70 pp.

Pasca, D., 1994. O garimpo de ouro de Poconé (MT): Estudos das inter-relações entre técnicas de extração, organização sócio-econômica e impactos ambientais. Monografia de Conclusão de Curso. Universidade de Tübingen, Tübingen, 68 pp.

Peixoto, J.L.S., Bezerra, M.A.O., Isquerdo, S.W.G., 1999. Padrão de assentamento das populaces indigenas pré – históricas do Pantanal Sul-Mato-Grossense Pantanal. In: EMBRAPA (Ed.), Simpósio sobre Recursos Naturais e Sócio-econômicos do Panatnal: Manejo e Conservação, 2. Corumbá. EMBRAPA-SPI, Corumbá, 1996, pp. 431–436.

Ponce, V.M., 1995. Hydrological and Environmental Impact of the Paraná-Paraguay Waterway on the Pantanal of Mato Grosso (Brazil). San Diego State University, San Diego, 125 pp.

Remppis, M., 1995. Fazendas zwischen Tradition und Fortschritt – Umweltauswirkungen der Rinderweidewirtschaft im nördlichen Pantanal. In: Kohlhepp, G. (Ed.), Mensch-Umwelt-Beziehungen in der Pantanal Region von Mato Grosso/Brasilien. Beiträge zur angewandten geographischen Umweltforschung. Tübinger Geographische Studien 114, 1–29.

Short, N.M., Blair Jr., R.W., 1986. Geomorphology from space. A global overview of regional landforms. NASA SP-486. Washington, DC. Available online at http://daac.gsfc.nasa.gov/daac_docs/geomorphology/geo_home_page.html.

Stevaux, J.C., 2000. Climatic events during the late Pleistocene and Holocene in the upper Parana River: Correlation with NE Argentina and South-Central Brazil. Quaternary Int. 72, 73–86.

The Nature Conservancy (TNC), 1996. Troubled waters: protecting our aquatic heritage. The Nature Conservancy, Arlington, VA, 17 pp.

Wantzen, K.M., Da Silva, C.J., Figueiredo, D.M., Migliácio, M.C., 1999. Recent impacts of navigation on the upper Paraguay. Rev. Bol. Ecol. 6, 173–182.

Willink, P.W., Chernoff, B., Alonso, L.E., Montanbault, J.R., Lourival, R., 2000. A biological assessment of the aquatic ecosystems of the Pantanal, Mato Grosso do Sul, Brazil. RAP Bulletin of Biological Assessment 18. Conservation International, Washington, DC, 305 pp.

Zeilhofer, P., 1996. Geoökologische Charakterisierung des nördlichen Pantanal von Mato Grosso, Brasilien, anhand multitemporaler Landsat Thematic Mapper-Daten. PhD Thesis. Herbert Utz Verlag, München, 225 pp.

Available online at www.sciencedirect.com

SCIENCE DIRECT°

Ecological Engineering 24 (2005) 403–418

**ECOLOGICAL
ENGINEERING**

www.elsevier.com/locate/ecoleng

ELSEVIER

Ecological engineering for successful management and restoration of mangrove forests

Roy R. Lewis III*

Lewis Environmental Services, Inc., P.O. Box 5430, Salt Springs, FL 32134, USA

Received 2 January 2004; received in revised form 22 September 2004; accepted 29 October 2004

Abstract

Great potential exists to reverse the loss of mangrove forests worldwide through the application of basic principles of ecological restoration using ecological engineering approaches, including careful cost evaluations prior to design and construction. Previous documented attempts to restore mangroves, where successful, have largely concentrated on creation of plantations of mangroves consisting of just a few species, and targeted for harvesting as wood products, or temporarily used to collect eroded soil and raise intertidal areas to usable terrestrial agricultural uses. I document here the importance of assessing the existing hydrology of natural extant mangrove ecosystems, and applying this knowledge to first protect existing mangroves, and second to achieve successful and cost-effective ecological restoration, if needed. Previous research has documented the general principle that mangrove forests worldwide exist largely in a raised and sloped platform above mean sea level, and inundated at approximately 30%, or less of the time by tidal waters. More frequent flooding causes stress and death of these tree species. Prevention of such damage requires application of the same understanding of mangrove hydrology.
© 2005 Elsevier B.V. All rights reserved.

Keywords: Mangrove forests; Restoration of mangrove forests; Ecological restoration; Mangroves

1. Introduction

Mangrove forests are ecologically important coastal ecosystems (Lugo and Snedaker, 1974) composed of one or more of the 69 species of plants called mangroves (Duke, 1992). These ecosystems currently cover 146,530 km of the tropical shorelines of the world (FAO, 2003). This represents a decline from 198,000 km of mangroves in 1980, and 157,630 km in 1990 (FAO, 2003). These losses represent about 2% per year between 1980 and 1990, and 1% per year between 1990 and 2000.

Examples of documented losses include combined losses in the Philippines, Thailand, Vietnam and Malaysia of 7445 km² of mangroves (Spalding, 1997). In Florida, approximately 2000 km² remain from an estimated historical cover of 2600 km² (Lewis et al., 1985). Puerto Rico has just 64 km² of mangrove remaining from an original mangrove forest cover estimated to have been 243 km² (Martinez et al., 1979). These figures emphasize the magnitude of the loss,

---

* Tel.: +1 352 546 4842; fax: +1 352 546 5224.
  E-mail address: lesrrl3@aol.com.

0925-8574/$ – see front matter © 2005 Elsevier B.V. All rights reserved.
doi:10.1016/j.ecoleng.2004.10.003

and the magnitude of the opportunities that exist to restore areas like mosquito control impoundments in Florida (Brockmeyer et al., 1997), and abandoned shrimp aquaculture ponds in Thailand and the Philippines (Stevenson et al., 1999), back to functional mangrove ecosystems.

Restoration of areas of damaged or destroyed mangrove forests has been previously discussed by Lewis (1982a,b, 1990a,b, 1994, 1999, 2000), Crewz and Lewis (1991), Cintron-Molero (1992), Field (1996, 1998), Turner and Lewis (1997), Brockmeyer et al. (1997), Milano (1999), Ellison (2000), Lewis and Streever (2000) and Saenger (2002). Saenger and Siddiqi (1993) describe the largest mangrove afforestation program in the world, with plantings of primarily one species (*Sonneratia apetala*) over 1600 km^2 on newly accreting mud flats in Bangladesh. This was a multipurpose planting with the prime objective of "... providing land sufficiently raised and stabilized to be used for agricultural purposes ..." through encouraged accretion of sediments by the plantings. It is estimated that 600 km^2 of raised lands have now been converted to such uses. Blasco et al. (2001) estimate survival of these plantings to presently cover about 800 km^2 after about a 50% loss due to cyclones and insect pest outbreaks.

In spite of the success in Bangladesh, most attempts to restore mangroves often fail completely, or fail to achieve the stated goals (Lewis, 1990a, 1999, 2000; Erftemeijer and Lewis, 2000). This paper is intended to review those factors that can be applied by ecological engineers and ecologists to insure successful management without damage, and successful restoration if damage has or does occur. In addition, following the suggestions in Weinstein et al. (2001), emerging restoration principles will be stated.

2. Key terms and principles

Restoration or rehabilitation may be recommended when an ecosystem has been altered to such an extent that it can no longer self-correct or self-renew. Under such conditions, ecosystem homeostasis has been permanently stopped and the normal processes of secondary succession (Clements, 1929) or natural recovery from damage are inhibited in some way. This concept has not been analyzed or discussed with any great detail as it pertains to mangrove forests (Detweiler et al., 1975; Ball, 1980; Lewis, 1982a,b, are the few exceptions), and thus restoration has, unfortunately, emphasized planting mangroves as the primary tool in restoration, rather than first assessing the reasons for the loss of mangroves in an area and working with the natural recovery processes that all ecosystems have.

The term "restoration" has been adopted here to specifically mean any process that aims to return a system to a pre-existing condition (whether or not this was pristine) (sensu Lewis, 1990c), and includes "natural restoration" or "recovery" following basic principles of secondary succession. Secondary succession depends upon mangrove propagule availability, and I suggest a new term, "propagule limitation" to describe situations in which mangrove propagules may be limited in natural availability due to removal of mangroves by development, or hydrologic restrictions or blockages (i.e. dikes) which prevent natural waterborne transport of mangrove propagules to a restoration site. Such situations have been described by Lewis (1979) for the U.S. Virgin Islands, Das et al. (1997) for a mangrove restoration site in the Mahanadi delta, Orissa, India, and by Hong (2000) for similar efforts at Can Gio, Vietnam.

"Ecological restoration" is another important term to include in this discussion and has been defined by the Society for Ecological Restoration (SER, 2002) as the "process of assisting the recovery of an ecosystem that has been degraded, damaged, or destroyed". The goal of this process is to emulate the structure, functioning, diversity and dynamics of the specified ecosystem using reference ecosystems as models.

Ecological engineering, which involves creating and restoring sustainable ecosystems that have value to both humans and nature (Mitsch and Jørgensen, 2004) has been characterized as having two primary goals: (1) the restoration of ecosystems that have been substantially disturbed by human activities ... and (2) the development of new sustainable ecosystems that have both human and ecological value, to which I would add a third, which is to accomplish items (1) and (2) in a cost effective way. Engineers are routinely asked to generate engineer's estimates for construction projects, often oversee actual construction, and approve payments based upon successful completion of construction. Associated materials purchase and installation, such as plants in a wetland restoration project, are other items reviewed, approved and paid for. Projected costs are

important to determine if a project is affordable, and final costs have to be controlled in the construction process.

As noted by Spurgeon (1999) "[I]f coastal habitat rehabilitation/creation is to be widely implemented, greater attempts should be made to: find ways of reducing the overall costs of such initiatives; devise means of increasing the rate at which environmental benefits accrue; and to identify mechanisms for appropriating the environmental benefits". It is the role of an ecological engineer, working in tandem with an ecologist, to see that such actions occur.

3. Ecology of mangrove forests

Mangroves are intertidal trees found along tropical shorelines around the world. They are frequently inundated by the tides, and thus have special physiological adaptations to deal with salt in their tissues. They also have adaptations within their root systems to support themselves in soft mud sediments and transport oxygen from the atmosphere to their roots, which are largely in anaerobic sediments. Most have floating seeds that are produced annually in large numbers and float to new sites for colonization.

Mangrove forests provide a number of ecological benefits including stabilizing shorelines, reducing wave and wind energy against shorelines, and thus protecting inland structures, supporting coastal fisheries for fish and shellfish through direct and indirect food support and provisions for habitat, and support of wildlife populations including a number of wading birds and sea birds.

Mangrove forests also support timber production for construction materials and supply some special chemicals for industry, and medicinal products for local use.

4. Ecological management of mangroves

As noted by Field (1998), "[T]he most common method of conserving mangrove ecosystems is by the creation of protected areas in undisturbed sites ..." National parks, wildlife preserves and internationally protected sites are mentioned. However, as reported by Perdomo et al. (1998), 70% of the Cienaga Grande de Santa Marta, a 511 km^2 mangrove forest reserve in

Colombia, have been killed by alterations of hydrology due to road and dike construction in the 1950s. Similar deaths of mangroves in a protected area due to modified hydrology are reported in Turner and Lewis (1997). Rubin et al. (1999) describe the destruction of the mangrove forests of the Volta River Estuary in Ghana due to two dams on the Volta River, and local timber harvesting. Ellison (2000) notes that "[D]espite repeated claims that mangrove forests can be managed sustainably ... managed (and unmanaged) mangal continues to degrade and disappear at rates comparable to those seen in tropical wet forests (~1.5% per year) ..."

Clearly, mangrove forests have not been managed very well, even if left alone in terms of direct dredging and filling for coastal development (Lewis, 1977), or conversion to aquaculture ponds (Stevenson et al., 1999). In case, after case disruption of the existing hydrology of a forest is enough to kill it. One might assume that all of these cases involved the old misunderstanding that mangroves were worthless swamps, and today we know how to manage them better. The example of Clam Bay in Naples, FL, USA, however, (Turner and Lewis, 1997) shows that even modern day management ignores the realities of mangrove hydrology.

The issue appears to be that both ecologists and engineers (and ecological engineers) do not understand mangrove hydrology. Although a number of papers discuss the science of mangrove hydrology (Kjerfve, 1990; Wolanski et al., 1992; Furukawa et al., 1997), their focus has been on tidal and freshwater flows within the forests, and not the critical periods of inundation and dryness that govern the health of the forest. Kjerfve (1990) does discuss the importance of topography and argues that "... micro-topography controls the distribution of mangroves, and physical processes play a dominant role in formation and functional maintenance of mangrove ecosystems ...". Hypersalinty due to year to year variations in rainfall can produce natural mangrove die-backs (Cintron et al., 1978), and disruption of normal freshwater flows that dilute seawater in more arid areas can kill mangroves (Perdomo et al., 1998; Medina et al., 2001). What is less understood is the role of tidal inundation frequency, and modifications to that factor, that can also stress and kill mangroves.

A series of papers beginning with Nickerson and Thibodeau (1985) and Thibodeau and Nickerson (1986), and continuing with McKee and Mendelssohn

(1988), McKee (1993, 1995a,b), and McKee and Faulkner (2000a,b) have clearly shown that differential survival and growth of mangrove species studied to date are related to the depth, duration and frequency of flooding and soil saturation. The processes involved are complicated and no single factor applies to all mangrove zones, but observations and data collection across transects through mangroves from low to higher elevations in Belize "... indicate that the higher-elevations sites were infrequently flooded over the soil surface, whereas the lower elevation sites near the shoreline were inundated twice daily. Tidal amplitude and water velocity decrease strongly with increasing distance from the shoreline and lead to restricted water movement and incomplete drainage of interior areas ...". In examining the correlations of measured environmental variables across transects with different dominant species of mangroves, three factors were examined for correlations with mangrove zonation. Within the three factors, flooding "had a high negative loading of relative elevation and a high positive loading of sulfide. Sulfide tends to accumulate in waterlogged soils, a process that is promoted in low elevation areas where water levels may not fall below the soil surface during a tidal cycle ...".

As noted by Koch et al. (1990) "sulfide toxicity has been implicated as a causative factor in the dieback of European and North American salt marshes ..." and Mendelssohn and Morris (2000) in reporting on the ecophysiological controls on the productivity of smooth cordgrass further define the toxic effects of sulfide as reducing ammonium uptake that "result in a plant nitrogen deficiency and lower rates of growth and primary production for poorly drained, inland *Spartina* marshes". A similar effect is likely in mangrove forests.

The point of all of this is that flooding depth, duration and frequency are critical factors in the survival of both mangrove seedlings and mature trees. Once established, mangroves can be further stressed if the tidal hydrology is changed, for example by diking (Brockmeyer et al., 1997). Both increased salinity due to reductions in freshwater availability, and flooding stress, increased anaerobic conditions and free sulfide availability can kill existing stands of mangroves.

For these reasons, any engineering works constructed near mangrove forests, or in the watershed that drains to mangrove forests, must be designed to allow for sufficient free exchange of seawater with the adjacent ocean or estuary, and not interrupt essential upland or riverine drainage into the mangrove forest. Failure to properly account for these essential inputs and exchange of water will result in stress and possible death of the forest.

5. Ecological restoration of mangroves

It has been reported that mangrove forests around the world can self-repair or successfully undergo secondary succession over periods of 15–30 years if: (1) the normal tidal hydrology has not been disrupted and (2) the availability of waterborne seeds or seedlings (propagules) of mangroves from adjacent stands is not limited or blocked (Lewis, 1982a; Cintron-Molero, 1992; Field, 1998).

Ecological restoration of mangrove forests has only received attention very recently (Lewis, 1999). The wide range of types of projects previously considered to be restoration, as outlined in Field (1996, 1998), reflect the many aims of classic mangrove rehabilitation or management for direct natural resource production. These include planting monospecific stands of mangroves for future harvest as wood products. This is not ecological restoration as defined above.

It is important to understand that mangrove forests occur in a wide variety of hydrologic and climatic conditions that result in a broad array of mangrove community types. In Florida, Lewis et al. (1985) have identified at least four variations on the original classic mangrove zonation pattern described by Davis (1940), all of which include a tidal marsh component dominated by such species as smooth cordgrass (*Spartina alterniflora*) or saltwort (*Batis maritima*). Lewis (1982a,b) describes the role that smooth cordgrass plays as a "nurse species", where it initially establishes on bare soil and facilitates primary or secondary succession to a climax community of predominantly mangroves, but with some remnant of the original tidal marsh species remaining. This has been further generalized by Crewz and Lewis (1991) (Fig. 1) as the typical mangrove forest for Florida, where tidal marsh components are nearly always present.

Finn (1996, 1999) describes the construction and operation of a mixed estuarine mesocosm as part of the Biosphere 2 experiment. Several of the subunits within the mesocosm contained mangroves transplanted from

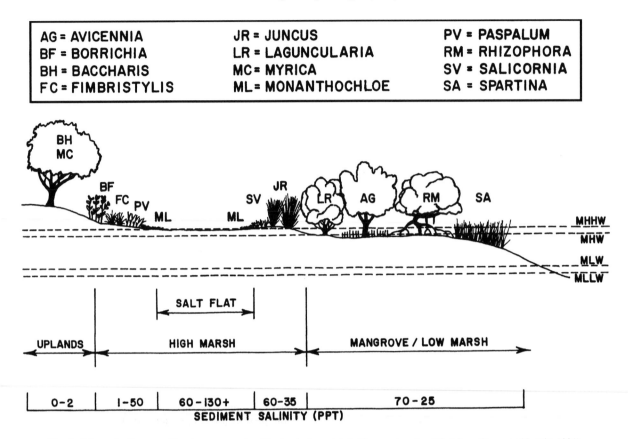

Fig. 1. Schematic diagram of the six components of the tropical coastal shelf ecosystem (modified from Crewz and Lewis, 1991).

Florida. No specific measurements of tidal inundation depth, duration and frequency at the source site of the mangroves were made, and the initial management of tidal effects in the mesocosm are not described in detail. The mesocosm and adjacent mesocosms exchanged water to simulate tides, but this was discontinued, and Finn (1996) indicates that the mangrove mesocosm had operated for 3 years without tides. The amount of inundation is not described in the non-tidal mesocosm, but Finn (1996) states that the experiment may be a useful tool for characterizing the effect of impounding mangroves. Finn (1999) describes the lack of understory vegetation in the mesocosm and notes that this compares favorably with natural systems. The transplanted mangroves have grown well in the mesocosm but most of the animals in the system, including fiddler crabs, periwinkles and coffee snails disappeared from the system between 1991 and 1993. There were restocked in 1994 but their fate is not reported in Finn (1999).

It is possible to restore some of the functions of a mangrove forest, salt flat or other systems even though parameters such as soil type and condition may have altered and the flora and fauna may have changed (Lewis, 1992). If the goal is to return an area to a pristine pre-development condition, then the likelihood of failure is increased. However, the restoration of certain ecosystem traits and the replication of natural functions stand more chance of success (Lewis et al., 1995).

Because mangrove forests may recover without active restoration efforts, it has been recommended that restoration planning should first look at the potential existence of stresses such as blocked tidal inundation that might prevent secondary succession from occurring, and plan on removing that stress before attempting restoration (Hamilton and Snedaker, 1984; Cintron-Molero, 1992). The next step is to determine by observation if natural seedling recruitment is occurring once the stress has been removed. Only if natural recovery is not occurring should the final step of considering

assisting natural recovery through planting be considered.

Unfortunately, many mangrove restoration projects move immediately into planting of mangroves without determining why natural recovery has not occurred. There may even be a large capital investment in growing mangrove seedlings in a nursery before stress factors are assessed. This often results in major failures of planting efforts. For example, Sanyal (1998) has recently reported that between 1989 and 1995, 9050 ha of mangroves were planted in West Bengal, India, with only a 1.52% success rate. In the Philippines, the Central Visayas Regional Project I, Nearshore Fisheries Component, a US$ 35 million World Bank Project targeted 1000 ha of mangrove planting between 1984 and 1992. An evaluation of the success of the planting in 1995–1996 by Silliman University (Silliman University, 1996; de Leon and White, 1999) indicated that only 18.4% of the 2,927,400 mangroves planted over 492 ha had survived. Another planned 30,000 ha planting effort funded by a US$ 150 million loan from the Asian Development Bank and Overseas Economic Cooperation Fund of Japan (Fisheries Sector Program, 1990–1995) was cut short after only 4792 ha were planted do to similar problems (Ablaza-Baluyut, 1995).

Platong (1998) in reporting on efforts at mangrove restoration in Thailand states that the Royal Forest Department of Thailand (RFD) reported 11,009 ha planted in Southern Thailand. Platong (1998) notes that RFD "is unable to justify the success of the plan because the replanted mangrove areas are just in seedling stage. There is no report that replanting mangroves are survived [sic] or destroyed by natural factors and human. The data being recorded are only the planted area and the amount of areas planned to be replanted" (p. 59). In addition "the Agriculture Department joined with the private sector in a mangrove replanting project for the King's 50th anniversary jubilee . . .. The target was 31,724 rai [5076 ha] in 57 areas. The Petrolium [sic] Authority of Thailand (PTT) replanted mangrove forest in Southern Thailand . . . between 1995 and 1997 about 11,062 rai [1770 ha] . . .. It is not easy to compare the success of mangrove replanting . . . because they are not the same scale, e.g. species, number of areas, location, timing and budget for maintenance after replanting". Platong (1998) also refers to planting of mangrove seeds or seedlings in areas that have not previously been forested.

Many of these failures result from afforestation attempts, which are an attempt to plant mangroves in areas that previously did not support mangroves. Often mudflats in front of existing or historical stands of mangroves are proposed restoration sites. Aside from the problem of frequent flooding greater than the tolerance of mangroves, it is questionable whether the widespread attempts to convert existing natural mudflats to mangrove forests, even if they succeeded, represent ecological restoration. In their review article on this matter, Erftemeijer and Lewis (2000) have commented that planting mangroves on mudflats would represent habitat conversion rather than habitat restoration, and strongly caution against the ecological wisdom of doing this.

Similar efforts in the Philippines, as reported by Custodio (1996), under "Threats to Shorebirds and their Habitats", state that "{H}abitat alteration in the wake of unabated increase in human population is still the most important threat to shorebirds in the Philippines. Some of the alteration, however, has been due to activities, which were of good intention. An example of this is the mangrove 'reforestation' programme which covered the feeding grounds of shorebirds in Puerto Rivas (Bataan) and parts of Olango Island" (p. 166). With these words in mind, it is worthwhile to note that Tunhikorn and Round (1996) state that ". . . Thailand is a major wintering and passage area for Palaeartic waterbirds. Large numbers of shorebirds are found both along its coastline, in mudflat and mangrove habitat . . ." and describe the intertidal mudflats, onshore prawn ponds, salt-pans and some remaining areas of mangroves along the Gulf of Thailand as "(P)robably the single most important site for shorebirds in the country" (p. 123). Finally, they describe the major threat to wintering shorebirds at Khao Sam Roi Yot National Park in Prachuap Khiri Khan province as modifications to "the hydrology and topography of coastal areas . . . by intensive prawn farming during 1988–1993" (p. 124).

Natural recruitment of mangrove seedlings, reflected in the careful data collection of Duke (1996) at an oil spill site in Panama showed that ". . . densities of *natural recruits* far exceeded both expected and observed densities of planted seedlings in both sheltered and exposed sites" (emphasis added) in restoration attempts at a previously oiled mangrove forest. Soemodihardjo et al. (1996) report that only 10% of a logged area in Tembilahan, Indonesia (715 ha) needed

replanting because "The rest of the logged over area ... had more than 2500 *natural seedlings* per ha" (emphasis added).

Lewis and Marshall (1997) have suggested five critical steps are necessary to achieve successful mangrove restoration:

1. Understand the autecology (individual species ecology) of the mangrove species at the site, in particular the patterns of reproduction, propagule distribution and successful seedling establishment.
2. Understand the normal hydrologic patterns that control the distribution and successful establishment and growth of targeted mangrove species.
3. Assess the modifications of the previous mangrove environment that occurred that currently prevents natural secondary succession.
4. Design the restoration program to initially restore the appropriate hydrology and utilize natural volunteer mangrove propagule recruitment for plant establishment.
5. Only utilize actual planting of propagules, collected seedlings or cultivated seedlings after determining through Steps 1–4 that natural recruitment will not provide the quantity of successfully established seedlings, rate of stabilization or rate of growth of saplings established as goals for the restoration project.

Callaway (2001) lists seven similar steps in order to design the best hydrology and geomorphological development of tidal marshes in California.

These critical steps are often ignored and failure in most restoration projects can be traced to proceeding in the early stages directly to Step 5, without considering Steps 1–4. Stevenson et al. (1999) refer to this approach as "gardening", where simply planting mangroves is seen as all that is needed. The successful plantings of large areas with one or two species, as described by Saenger and Siddiqi (1993), in Bangladesh, may seem a success story, but one must question whether large monotypic stands of mangroves are a worthwhile goal. Remembering the principles of ecological restoration, one should ask whether the results produce a mangrove forest similar in species composition and faunal use to the native mangrove forests of the area. Another issue is competition from large-scale plantings may prevent natural colonization by volunteer mangroves, and reduce the final biodiversity of the planted area. Another

common problem is the failure to understand the natural processes of secondary succession, and the value of utilizing nurse species like smooth cordgrass in situations where wave energy may be a problem.

As an example of the problem, Kairo et al. (2001) in a recent paper with a title similar to this paper begin their section on "[H]istory of mangrove restoration and management" with this statement: "[M]angrove *planting* and management has a long history ..." (emphasis added). Spurgeon (1999) does the same thing. Under his section on "Costs", for mangrove rehabilitation/creation it begins "[C]osts for mangrove *planting* can range ..." (emphasis added). Although Kairo et al. (2001) later have a section on "natural regeneration" the emphasis throughout their paper is on planting. Thus, for the majority of papers written on mangrove restoration, there is an immediate assumption that mangrove restoration means mangrove planting. This leads then to ignoring hydrology and natural regeneration via volunteer mangrove propagules, and many failures in attempts to restore mangroves (Erftemeijer and Lewis, 2000).

The single most important factor in designing a successful mangrove restoration project is determining the normal hydrology (depth, duration and frequency, and of tidal flooding) of existing natural mangrove plant communities (a reference site) in the area in which you wish to do restoration. Both Vivian-Smith (2001) and Sullivan (2001), similarly recommend the use of a reference tidal marsh for restoration planning and design. The normal surrogate for costly tidal data gathering or modeling is the use of a tidal benchmark and survey of existing healthy mangroves. When this is done, a diagram similar to that in Fig. 1 will result. This then becomes the construction model for your project.

Fig. 1 is a typical cross section through a reference mangrove forest site. Actual survey data is generated to locate the existing topographic elevations within the forest. This figure is a synthesis of all the topographic information generated by Crewz and Lewis (1991). Table 1 modified from Detweiler et al. (1975) is actual data from a single mangrove forest on Tampa Bay, Florida. Both Fig. 1 and Table 1 show that the mangrove forests in Florida typically exist on a sloped platform above mean sea level, with typical surveyed elevations for mangrove species in the range of +30 to +60 cm above mean sea level. Likewise, Twilley and Chen (1998) report the topography of a basin mangrove

Table 1
Elevation ranges and mean elevation (NGVD datum) of 10 plant species found in the control transect of an undisturbed mangrove forest community near Wolf Branch Creek, Tampa Bay, FL, USA (modified from Detweiler et al., 1975)

Species	Number of quadrats	Range (ft)	Range (m)	Mean elevation (ft)	Mean elevation (m)
Rhizophora mangle	35	+0.2 to +1.6	+0.06 to +0.49	+1.0	+0.30
Avicennia germinans	49	+0.4 to +2.5	+0.12 to +0.76	+1.5	+0.46
Laguncularia racemosa	47	+0.7 to +2.5	+0.21 to +0.76	+1.5	+0.46
Spartina alterniflora	4	+1.6 to +1.7	+0.49 to +0.52	+1.7	+0.52
Salicornia virginica	10	+1.6 to +1.9	+0.49 to +0.58	+1.7	+0.52
Sesuvium portulcastrum	2	+1.7	+0.52	+1.7	+0.52
Limonium carolinianum	6	+1.6 to +1.7	+0.49 to +0.52	+1.7	+0.52
Batis maritima	14	+1.6 to +2.2	+0.49 to +0.67	+1.8	+0.55
Borrichia frutescens	2	+1.9	+0.58	+1.9	+0.58
Philoxerus vermicularis	5	+1.6 to +2.2	+0.49 to +0.67	+1.9	+0.58

forest at Rookery Bay had a "... bowl shape with a centre low of 45 cm > msl". A similar profile section from Whitten et al. (1987) for a different group of mangrove species in Sumatra shows a similar pattern (Fig. 2). Finally, in Fig. 3, four sites in Australia are illustrated from Kenneally (1982). All show a similar location, at the upper third of the tidal range. Kjerfve (1990) reports that within the Klong Ngao creek-mangrove system in Thailand "... the mangrove wetland area above bank-full stage is only inundated 9% of the time. Specific locations within the wetland at higher elevations are flooded less frequently, and the system as a whole is only inundated 1% of the time".

In an early review of percent tidal submergence and emergence for tidal marshes, Hinde (1954) reported that the tidal marsh in Palo Alto, California, had zones of tidal marsh vegetation that varied in their percent of time submerged from 20% for the highest *Salicornia*

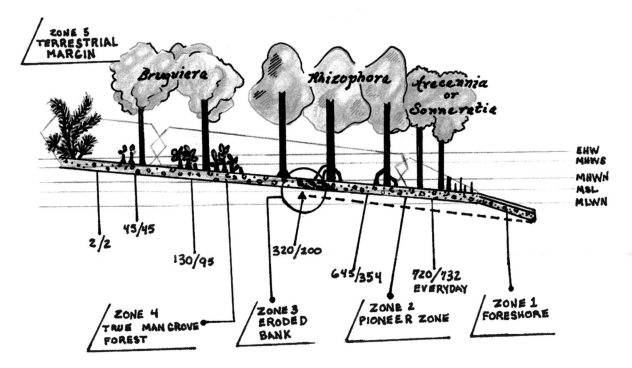

Fig. 2. Mangrove zonation related to tidal datums in Sumatra, Indonesia (modified from Whitten et al., 1987).

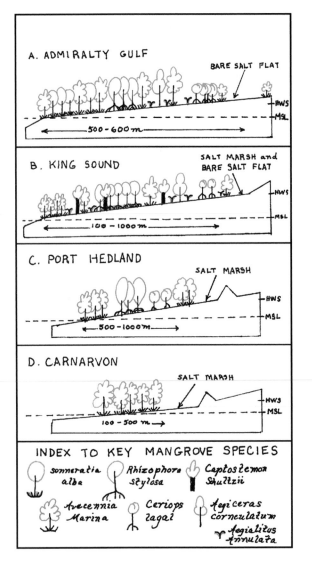

Fig. 3. Zonation of mangroves in western Australia (modified from Kenneally, 1982). Line added to emphasize mean sea level datum.

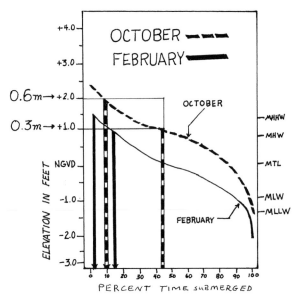

Fig. 4. The topographic position of mangroves on Tampa Bay, FL, USA (i.e. +0.3 to +0.6 m) in relationship to the percent time of submergence (modified from Lewis and Estevez, 1988).

to 80% for the lowest *Spartina*. Thus, tidal marshes appear to have a range of tolerance for submergence greater than that of mangrove forests.

The implications of these data are significant, and often overlooked. First, it appears, based on the data generated to date that mangrove forests around the world have a similar pattern of occurrence, regardless of species composition, on a tidal plane above mean high water and extending to high water spring elevations. Second, this means that the time during which

mangroves are typically inundated by high tides is very restricted. Figs. 4 and 5 show two illustrations of the actual period of time that mangrove forests on Tampa Bay, FL, USA (Fig. 4 from Lewis and Estevez, 1988) and Gladstone, Queensland (Fig. 5 from Hutchings and Saenger, 1987) are inundated with tidal waters. Both sets of inundation curves relative to topography show that total time of inundation throughout a typical year is 30% or less. Fig. 4 shows the topographic zone within which mangroves occur on Tampa Bay (+0.3 to +0.6 m) and how frequently that zone is likely to be flooded based upon tide curves. Detailed studies of the Rookery Bay mangroves (Twilley and Chen, 1998) show similar data, with 152–158 tides per year recorded in two basin mangrove forests out of a potential of 700+ high tides per year in a system with mixed diurnal tides. Cahoon and Lynch (1997) report data for continuous water level monitoring in three red mangrove (*Rhizophora mangle*) forests, and one basin forest in southwest Florida. The mean total hours of flooding over a 2-year-period for the red mangrove forest was 6055 or 35.3% of the potential total for the three sites. The mean number of flooding events was 1184 or 1.65 tides per day. In contrast, the single basin forest site was flooded just 88 times in 2 years, yet total hours of flooding were 10,182

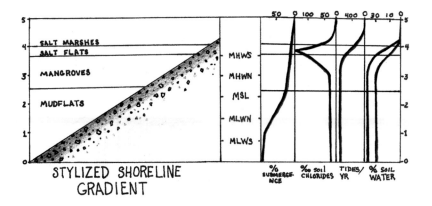

Fig. 5. Integration of vegetational boundaries with gradient-related and tidally induced boundary conditions based on data collected from study areas in Gladstone, Queensland, 1975–1983 (modified from Hutchings and Saenger, 1987).

or 59.4% of the potential time reflecting the trapping of both tidal waters and rainfall. This is not the prevailing understanding of mangrove tidal hydrology.

For example, Watson (1928) created five inundation classes ranging from Class 1, "inundated by all high tides", to Class 5, occasionally inundated by exceptional or equinoctial tides", and placed all the mangroves at his location in Malaysia in Classes 2–5 with distinct zonation based upon the nature of the tide that inundates an area rather than the number of times or total period of inundation. Field (1998) makes reference to topographical and hydrological changes to mangrove sites as a key to understanding rehabilitation needs, but provides no specific information. Perdomo et al. (1998) states that "[M]angroves may grow at sites which are permanently covered by shallow water ..." without providing data to support this statement.

Although many authors note that mangroves appear to be limited to certain ground elevations relative to flooding frequency (Watson, 1928; Field, 1996; Ellison, 2000), few have ever quantified it, as noted above, and fewer still recognize the importance of this issue relative to mangrove management and restoration.

Options for restoration, as discussed before, include simply restoring hydrologic connections to impounded mangroves (Brockmeyer et al., 1997). Another is the construction, by excavation of fill or backfilling of an excavated area, to create a target restoration site with the same general slope, and the exact tidal elevations relative to a benchmark as the reference site, thus insuring that the hydrology is correct. The final graded topography of a site needs to be designed to match that

found in an adjacent reference forest and checked carefully by survey during and at the completion of construction. Crewz and Lewis (1991) in examining the critical issues in success and failure in tidal marsh and mangrove restoration in Florida found that the hydrology, as created or restored by excavation to the correct tidal elevation, was the single most important element in project success. This is similar to the recommendations of Rozas and Zimmerman (1994) (as cited in Streever, 2000) for smooth cordgrass marsh creation on dredged material. Similar focused attention to the topographic grade relative to adjacent natural mangroves in constructed mangrove wetlands was shown to be the key to success in a project at Brisbane International Airport in Australia (Saenger, 1996).

McKee and Faulkner (2000a) report that two mangrove restoration sites were constructed respectively to grades of +45 cm (Site WS) and +43 cm (Site HC) relative to National Vertical Geodetic Datum (NGVD). No mention is made of how these elevations were determined. One of the referenced sites (WS) is described by Stephen (1984) as actually having variable final topographic elevations ranging from +24 cm to +190 cm at the time of completion of construction, with the +45 cm elevation being the original target elevation based upon surveys of the surrounding mature mangroves. Stephen (1984) noted that the best observed growth of mangroves was at +39 cm. Both Stephen (1984) and McKee and Faulkner (2000a) suggested the value of creating tidal creeks as part of these mangrove restoration projects in order to improve flushing. This is a predominant theme also in Zedler (2001) related to

Fig. 6. Time series photographs of a hydrologic mangrove restoration project at West Lake Park, Hollywood, FL, USA (A) Time Zero, July 1989, (B) Time Zero + 28 months, November 1991 and (C) Time Zero + 78 months, January 1996. No planting of mangroves occurred. All vegetation derived from volunteer mangrove propagules.

tidal marsh restoration. Stephen (1984) also notes that consideration should be given to intentional variation of grade and creation of permanent ponded areas to provide habitat for small fish, wading birds, algae and oysters.

Fig. 6A–C show a time sequence over a period of 78 months from the completion of a portion of a hydrologic restoration at a 500 ha mangrove restoration site at West Lake near Fort Lauderdale, Florida. Lewis (1990a) describes the details of the work, but again success resulted from using a reference site, and targeting final constructed grades as the same as the adjacent undisturbed forest. This resulted in a final sloped grade from +27 cm to +42 cm MSL. Extensive constructed tidal creeks were also added to the original plans, which had been designed without them. No planting of mangroves took place or was necessary. All three of the Florida species of mangroves (red mangrove (*Rhizophora mangle*), black mangrove (*Avicennia germinans*) and white mangrove (*Laguncularia racemosa*). volunteered on their own. Another form of this hydrologic restoration is to reconnect impounded mangroves to normal tidal influence (Turner and Lewis, 1997; Brockmeyer et al., 1997).

Both of these typical options require detailed review and discussion between an ecological engineer and a mangrove restoration ecologist. Further inputs may be needed from a surveyor, hydrologist, a geologist and finally the client paying the bills.

6. Controlling the costs of restoration

Lewis (submitted for publication) reports that the range of reported costs for mangrove restoration was US$ 225–216,000 ha^{-1} without the cost of the land. Brockmeyer et al. (1997) was able to keep restoration costs to US$ 250 ha^{-1} with careful placement of culverted openings to impounded mangrove wetlands along the Indian River Lagoon, USA. Similar types of this hydrologic restoration are reported in Turner and Lewis (1997). Milano (1999) described in some detail the planning and construction process for ten wetland restoration projects in Biscayne Bay, FL, USA (Miami), of which eight were mangrove restoration projects. Careful planning to achieve success was emphasized, as were methods of insuring cost control. The eight projects ranged in cost

from US$ 4286–214,285 ha^{-1}, with a mean of US$ 100,308 ha^{-1}. King (1998) has updated his 1993 cost estimates (King and Bohlen, 1994) to 1997 cost estimates for various wetland restoration costs and lists mangrove restoration at US$ 62,500 ha^{-1} excluding any land costs. Lewis Environmental and Coastal Environmental (1996) give cost estimates of US$ 62,500 ha^{-1} for government tidal wetland restoration attempts and US$ 125,000 ha^{-1} for private efforts, again without factoring in land costs. It is obvious that at these rates, mangrove restoration can be expensive, and therefore should be designed to be successful to avoid wasting large amounts of hard-to-get restoration dollars.

7. Emerging restoration principles

1. Get the hydrology right first.
2. Do not build a nursery, grow mangroves and just plant some area currently devoid of mangroves (like a convenient mudflat). There is a reason why mangroves are not already there or were not there in the recent past or have disappeared recently. Find out why.
3. Once you find out why, see if you can correct the conditions that currently prevent natural colonization of the selected mangrove restoration site. If you cannot correct those conditions, pick another site.
4. Use a reference mangrove site for examining normal hydrology for mangroves in your particular area. Either install tide gauges and measure the tidal hydrology of a reference mangrove forest or use the surveyed elevation of a reference mangrove forest floor as a surrogate for hydrology, and establish those same range of elevations at your restoration site or restore the same hydrology to an impounded mangrove by breaching the dikes in the right places. The "right places" are usually the mouths of historic tidal creeks. These are often visible in vertical (preferred) or oblique aerial photographs.
5. Remember that mangrove forests do not have flat floors. There are subtle topographic changes that control tidal flooding depth, duration and frequency. Understand the normal topography of your reference forest before attempting to restore another area.
6. Construction of tidal creeks within restored mangroves forests facilitates flooding and drainage, and allows for entree and exit of fish with the tides.
7. Evaluate costs of restoration early in project design to make your project as cost-effective as possible.

8. Conclusions

Ellison (2000) asks the question "mangrove restoration: do we know enough?" His answer is that "[R]estoration of mangal does not appear to be especially difficult ..." and comments that in contrast to the difficulties in restoring inland wetlands, "... it is more straightforward to restore tidal fluctuations and flushing to impounded coastal systems where mangroves could subsequently flourish ...". Thus, ecological restoration of mangrove forests is feasible, has been done on a large-scale in various parts of the world and can be done cost effectively. Lewis (2000) however, has pointed out that the failure to adequately train, and retrain coastal managers (including ecological engineers) in the basics of successful coastal habitat restoration all too often leads to projects "destined to fail, or only partially achieve their stated goals". The National Academy of Science of the United States in their report entitled "Restoring and Protecting Marine Habitat—The Role of Engineering and Technology" (National Research Council, 1994) stated that "the principle obstacles to wider use of coastal engineering capabilities in habitat protection, enhancement, restoration and creation are the cost and the institutional, regulatory and management barriers to using the best available technologies and practices" (emphasis added).

It is unfortunate that much of the research into mangrove restoration that has been carried out to date has been conducted without adequate site assessment, and without documentation of the methodologies or approaches used, and that it often lacks subsequent follow-up or evaluation. Unsuccessful (or only partially successful) projects are rarely documented. Field (1998) reports that after contacting numerous international organizations to get an overview of mangrove restoration work worldwide, "(T)he response was almost complete silence". He attributed this to bureaucratic sloth, proprietary reluctance to reveal important findings, inadequate dissemination mechanisms and a

myopic view of the general importance of rehabilitation programmes. I would add that few scientists or organizations wish to report or document failures.

In summary, a common ecological engineering approach should be applied to habitat restoration projects. The simple application of the five steps to successful mangrove restoration outlined by Lewis and Marshall (1997) would at least insure an analytical thought process and less use of "gardening" of mangroves as the solution to all mangrove restoration problems. Those involved could then begin to learn from successes or failures, act more effectively and spend limited mangrove restoration monies in a more cost-effective manner.

References

Ablaza-Baluyut, E., 1995. The Philippines fisheries sector program. In: Coastal and Marine Environmental Management. Proceedings of a workshop, 27–28 March 1995, Bangkok, Thailand, Asian Development Bank, Bangkok, Thailand, pp. 156–177.

Ball, M.C., 1980. Patterns of secondary succession in a mangrove forest in south Florida. Oecologia (Berl.) 4, 226–235.

Blasco, F., Aizpuru, M., Gers, C., 2001. Depletion of the mangroves of continental Asia. Wetland Ecol. Manag. 9 (3), 245–256.

Brockmeyer Jr., R.E., Rey, J.R., Virnstein, R.W., Gilmore, R.G., Ernest, L., 1997. Rehabilitation of impounded estuarine wetlands by hydrologic reconnection to the Indian River Lagoon, Florida (USA). Wetland Ecol. Manag. 4 (2), 93–109.

Cahoon, D.R., Lynch, J.C., 1997. Vertical accretion and shallow subsidence in a mangrove forest of southwestern Florida, USA. Mangroves Salt Marshes 1 (3), 173–186.

Callaway, J.C., 2001. Hydrology and substrate. In: Zedler, J.B. (Ed.), Handbook for Restoring Tidal Wetlands. CRC Press, Boca Raton, Florida, pp. 89–117 (Chapter 3).

Cintron-Molero, G., 1992. Restoring mangrove systems. In: Thayer, G.W. (Ed.), Restoring the Nation's Marine Environment. Maryland Seagrant Program, College Park, Maryland, pp. 223–277.

Cintron, G., Lugo, A.E., Pool, D.J., Morris, G., 1978. Mangroves of arid environments in Puerto Rico and adjacent islands. Biotropica 10, 110–121.

Clements, F.E., 1929. Plant Competition. Carnegie Institute of Washington Publications, p. 398.

Crewz, D.W., Lewis, R.R., 1991. An Evaluation of Historical Attempts to Establish Emergent Vegetation in Marine Wetlands in Florida. Florida Sea Grant Technical Publication No. 60. Florida Sea Grant College, Gainesville, Florida.

Custodio, C.C., 1996. Conservation of migratory waterbirds and their wetland habitats in the Philippines. In: Wells, D.R., Mundikur, T. (Eds.), Conservation of migratory waterbirds and their wetland habitats in the East Asian-Australasian flyway. Proceedings of an International Workshop, Kushiro, Japan. 28 November–3 December, 1994. Wetlands International Asia Pacific, Kuala Lumpur, Publication No. 116, 163–173.

Davis, J.H., 1940. The ecology and geologic role of mangroves in Florida. Carnegie Inst. Wash. Pap. Tortugas Lab. No. 32. Publ. 517, 305–412.

Das, P., Basak, U.C., Das, A.B., 1997. Restoration of the mangrove vegetation in the Mahanadi Delta, Orissa, India. Mangroves Salt Marshes 1 (3), 155–161.

de Leon, T.O.D., White, A.T., 1999. Mangrove rehabilitation in the Philippines. In: Streever, W. (Ed.), An International Perspective on Wetland Rehabilitation. Kluwer Academic Publishers, The Netherlands, pp. 37–42.

Detweiler, T.E., Dunstan, F.M., Lewis, R.R., Fehring, W.K., 1975. Patterns of secondary succession in a mangrove community. In: Lewis, R.R. (Ed.), Proceedings of the Second Annual Conference on Restoration of Coastal Plant Communities in Florida. Hillsborough Community College, Tampa, Florida, pp. 52–81.

Duke, N., 1992. Mangrove floristics and biogeography. In: Robertson, A.I., Alongi, D.M. (Eds.), Tropical Mangrove Ecosystems. American Geophysical Union, Washington, DC, pp. 63–100.

Duke, N., 1996. Mangrove reforestation in Panama. In: Field, C. (Ed.), Restoration of Mangrove Ecosystems. International Society for Mangrove Ecosystems, Okinawa, Japan, pp. 209–232.

Ellison, A.M., 2000. Mangrove restoration: do we know enough? Rest. Ecol. 8 (3), 219–229.

Erftemeijer, P.L.A., Lewis, R.R., 2000. Planting mangroves on intertidal mudflats: habitat restoration or habitat conversion? In: Proceedings of the ECOTONE VIII Seminar Enhancing Coastal Ecosystems Restoration for the 21st Century, Ranong, Thailand, 23–28 May 1999. Royal Forest Department of Thailand, Bangkok, Thailand, pp. 156–165.

Food and Agricultural Organization (FAO), 2003. New global mangrove estimate. http://www.fao.org/forestry/foris/webview/forestry2/index.jsp%3Fgeold=0%26langid.

Field, C.D. (Ed.), 1996. Restoration of Mangrove Ecosystems. International Society for Mangrove Ecosystems, Okinawa, Japan.

Field, C.D., 1998. Rehabilitation of mangrove ecosystems: an overview. Mar. Pollut. Bull. 37 (8–12), 383–392.

Finn, M., 1996. The mangrove mesocosm of Biosphere 2: design, establishment and preliminary results. Ecol. Eng. 6, 21–56.

Finn, M., 1999. Mangrove ecosystem development in Biosphere 2. Ecol. Eng. 13, 173–178.

Furukawa, K.E., Wolanski, E., Mueller, H., 1997. Currents and sediment transport in mangrove forests. Estuar. Coast. Shelf Sci. 44, 301–310.

Hamilton, L.S., Snedaker, S.C. (Eds.), 1984. Handbook of Mangrove Area Management. East West Centre, Honolulu, Hawaii.

Hinde, H.P., 1954. The vertical distribution of salt marsh phanerograms in relation to tide levels. Ecol. Monogr. 24 (2), 210–225.

Hong, P.N., 2000. Effects of mangrove restoration and conservation on the biodiversity and environment in Can Gio district, Ho Chi Minh City. In: Asia-Pacific Cooperation on Research for Conservation of Mangroves, Proceedings of an International Workshop. The United Nations University, Tokyo, Japan, pp. 97–116.

Hutchings, P., Saenger, P., 1987. Ecology of Mangroves. University of Queensland Press, New York.

Kairo, J.G., Dahdouh-Guebas, F., Bosire, J., Koedam, N., 2001. Restoration and management of mangrove systems—a lesson from East African region. S. Afr. J. Bot. 67, 383–389.

Kenneally, K.F., 1982. Mangroves of Western Australia. In: Clough, B.F. (Ed.), Mangrove Ecosystems in Australia—Structure, Function and Management. Australian National University Press, Canberra, Australia, pp. 95–110.

King, D., 1998. The dollar value of wetlands: trap set, bait taken, don't swallow. Nat. Wetlands Newslett. 20 (4), 7–11.

King, D., Bohlen, C., 1994. Estimating the costs of restoration. Nat. Wetlands Newslett. 16 (3), 3-5+8.

Kjerfve, B., 1990. Manual for investigation of hydrological processes in mangrove ecosystems. UNESCO/UNDP Regional Project, Research and Its Application to the Management of the Mangroves of Asia and the Pacific (RAS/86/120).

Koch, M.S., Mendelssohn, I.A., McKee, K.L., 1990. Mechanism for the hydrogen sulfide-induced growth limitation in wetland macrophytes. Limnol. Ocean 35 (2), 399–408.

Lewis, R.R., 1977. Impacts of dredging in the Tampa Bay estuary 1876–1976. In: Pruitt, E.L. (Ed.), Proceedings of the Second Annual Conference of the Coastal Society—Time-stressed Environments: Assessment and Future Actions. The Coastal Society, Arlington, Virginia, pp. 31–55.

Lewis, R.R., 1979. Large scale mangrove restoration in St. Croix, U. S. Virgin Islands. In: Cole, D.P. (Ed.), Proceedings of the Sixth Annual Conference on Restoration and Creation of Wetlands. Hillsborough Community College, Tampa, Florida, pp. 231–242.

Lewis, R.R., 1982a. Mangrove forests. In: Lewis, R.R. (Ed.), Creation and Restoration of Coastal Plant Communities. CRC Press, Boca Raton, Florida, pp. 153–172.

Lewis, R.R., 1982b. Low marshes, peninsular Florida. In: Lewis, R.R. (Ed.), Creation and Restoration of Coastal Plant Communities. CRC Press, Boca Raton, Florida, pp. 147–152.

Lewis, R.R., 1990a. Creation and restoration of coastal plain wetlands in Florida. In: Kusler, J.A., Kentula, M.E. (Eds.), Wetland Creation and Restoration: The Status of the Science. Island Press, Washington, DC, pp. 73–101.

Lewis, R.R., 1990b. Creation and restoration of coastal wetlands in Puerto Rico and the US Virgin Islands. In: Kusler, J.A., Kentula, M.E. (Eds.), Wetland Creation and Restoration: The Status of the Science. Island Press, Washington, DC, pp. 103–123.

Lewis, R.R., 1990c. Wetlands restoration/creation/enhancement terminology: suggestions for standardization. In: Kusler, J.A., Kentula, M.E. (Eds.), Wetland Creation and Restoration: The Status of the Science. Island Press, Washington, D.C, pp. 417–422.

Lewis, R.R., 1992. Coastal habitat restoration as a fishery management tool. In: Stroud, R.H. (Ed.), Stemming the Tide of Coastal Fish Habitat Loss, Proceedings of a Symposium on Conservation of Coastal Fish Habitat, Baltimore, Md., 7–9 March 1991. National Coalition for Marine Conservation, Inc., Savannah, Georgia, pp. 169–173.

Lewis, R.R., 1994. Enhancement, restoration and creation of coastal wetlands. In: Kent, D.M. (Ed.), Applied Wetlands Science and Technology. Lewis Publishers, Inc., Boca Raton, Florida, pp. 167–191.

Lewis, R.R., 1999. Key concepts in successful ecological restoration of mangrove forests. In: Proceedings of the TCE-Workshop No.

II, Coastal Environmental Improvement in Mangrove/Wetland Ecosystems, 18–23 August 1998, Danish-SE Asian Collaboration on Tropical Coastal Ecosystems (TCE) Research and Training, Network of Aquaculture Centres in Asia-Pacific, Bangkok, Thailand, pp. 19–32.

Lewis, R.R., 2000. Ecologically based goal setting in mangrove forest and tidal marsh restoration in Florida. Ecol. Eng. 15 (3–4), 191–198.

Lewis, R.R., submitted for publication. Mangrove restoration—costs and benefits of successful ecological restoration. Proceedings of the Mangrove Valuation Workshop, Universiti Sans Malaysia, Penang, April 4–8, 2001. Beijer International Institute of Ecological Economics, Stockholm, Sweden.

Lewis, R.R., Gilmore Jr., R.G., Crewz, D.W., Odum, W.E., 1985. Mangrove habitat and fishery resources of Florida. In: Seaman, W. (Ed.), Florida Aquatic Habitat and Fishery Resources, Florida Chapter. American Fisheries Society, Eustis, Florida, pp. 281–336.

Lewis, R.R., Kusler, J.A., Erwin, K.L., 1995. Lessons learned from five decades of wetland restoration and creation in North America. In: Montes, C., Oliver, G., Molina, F., Cobos, J. (Eds.), In: Ecological Basis of Restoration of Wetlands in the Mediterranean Basin, University of La Rabida(Huelva) Spain, 7–11 June 1993 Junta de Andalucia, Spain, pp. 107–122.

Lewis, R. R., Marshall, M. J., 1997. Principles of successful restoration of shrimp Aquaculture ponds back to mangrove forests. Programa/resumes de Marcuba '97, September 15/20, Palacio de Convenciones de La Habana, Cuba, p. 126.

Lewis, R. R., Estevez, E. D., 1988. The Ecology of Tampa Bay, Florida: An Estuarine Profile. National Wetlands Research Center, US Fish and Wildlife Service, Biological Report No. 85 (7.18), Washington, DC.

Lewis, R. R., Streever, W., 2000. Restoration of Mangrove Habitat. Tech Note ERDC TN-WRP-VN-RS-3. US Army, Corps of Engineers, Waterways Experiment Station, Vicksburg, Mississippi.

Lewis Environmental Services, Inc., Coastal Environmental, Inc., 1996. Setting Priorities for Tampa Bay Habitat Protection and Restoration: Restoring the Balance. Tampa Bay National Estuary Program, Technical Publication #09-95, St. Petersburg, Florida.

Lugo, A.E., Snedaker, S.C., 1974. The ecology of mangroves. In: Johnson, R.F., Frank, P.W., Michener, C.D. (Eds.), Annual Review of Ecology and Systematics, 5, pp. 39–64.

Martinez, R., Cintron, G., Encarnacion, L.A., 1979. Mangroves in Puerto Rico: A Structural Inventory. Department of Natural Resources, Government of Puerto Rico, San Juan, Puerto Rico.

McKee, K.L., 1993. Soil physiochemical patterns and mangrove species distribution-reciprocal effects? J. Ecol. 81, 477–487.

McKee, K.L., 1995a. Seedling recruitment patterns in a Belizean mangrove forest: effects of establishment ability and physiochemical factors. Oecologia 101, 448–460.

McKee, K.L., 1995b. Interspecific variation in growth, biomass partitioning, and defensive characteristics of neotropical mangrove seedlings: response to availability of light and nutrients. Am. J. Bot. 82 (3), 299–307.

McKee, K.L., Mendelssohn, I.A., Hester, M.W., 1988. Reexamination of porewater sulfide concentrations and redox potentials near

the aerial roots of *Rhizophora mangle* and *Avicennia germinans*. Am. J. Bot. 75 (9), 1352–1359.

McKee, K.L., Faulkner, P.L., 2000a. Restoration of biogeochemical function in mangrove forests. Rest. Ecol. 8 (3), 247–259.

McKee, K.L., Faulkner, P.L., 2000b. Mangrove peat analysis and reconstruction of vegetation history at the Pelican Cays, Belize. Atoll Res. Bull. No. 46, 45-58.

Medina, E., Fonseca, H., Barboza, F., Francisco, M., 2001. Natural and man-induced changes in a tidal channel mangroves system under tropical semiarid climate at the entrance to the Maracaibo lake (Western Venezuela). Wetland Ecol. Manag. 9 (3), 233–243.

Mendelssohn, I.A., Morris, J.T., 2000. Ecophysiological controls on the productivity of *Spartina alterniflora* Loisel. In: Weinstein, M.P., Kreeger, D.A. (Eds.), Concepts and Controversies in Tidal Marsh Ecology. Kluwer Academic Publishers, Boston, pp. 59–80.

Milano, G.R., 1999. Restoration of coastal wetlands in southeastern Florida. Wetland J. 11 (2), 15–24, 29.

Mitsch, W.J., Jørgensen, S.E., 2004. Ecological Engineering and Ecosystem Restoration. John Wiley and Sons, Inc., Hoboken, NJ.

Nickerson, N.H., Thibodeau, F.R., 1985. Associations between pore water sulfide concentrations and the distribution of mangroves. Biogeochemistry 1, 183–192.

National Research Council, 1994. Restoring and Protecting Marine Habitat-The Role of Engineering and Technology. National Academy Press, Washington, DC.

Perdomo, L., Ensminger, I., Espinosa, L.F., Elster, C., Wallner-Kersanach, M., Schnetter, M.-L., 1998. The mangrove ecosystem of Cienaga Grande de Santa Marta (Colombia): observations on regeneration and trace metals in sediment. Mar. Pollut. Bull. 37 (8–12), 393–403.

Platong, J., 1998. Status of Mangrove Forests in Southern Thailand. Wetlands International—Thailand Programme. Hat Yai, Thailand, Publication No. 5.

Rozas, L.P., Zimmerman, R.J., 1994. Developing design parameters for constructing ecologically functional marshes using dredged material in Galveston Bay, Texas. In: Dredging'94, Proceedings of the second International Conference on Dredging and Dredged Material Placement, vol. 1. American Society of Civil Engineers, New York, NY, pp. 810–822.

Rubin, J.A., Gordon, C., Amatekpor, J.K., 1999. Causes and consequences of mangrove deforestation in the Volta Estuary, Ghana. Some recommendations for ecosystem restoration. Mar. Pollut. Bull. 37 (8–12), 441–449.

Saenger, P., 1996. Mangrove restoration in Australia: a case study of Brisbane International Airport. In: Field, C.D. (Ed.), Restoration of Mangrove Ecosystems. International Society for Mangrove Ecosystems, Okinawa, Japan, pp. 36–51.

Saenger, P., 2002. Mangrove Ecology. In: Silviculture and Conservation. Kluwer Academic Publishers, Dordrecht, The Netherlands.

Saenger, P., Siddiqi, N.A., 1993. Land from the seas: the mangrove afforestation program of Bangladesh. Ocean Coastal Manag. 20, 23–39.

Sanyal, P., 1998. Rehabilitation of degraded mangrove forests of the Sunderbans of India. In: Program of the International Workshop on Rehabilitation of Degraded Coastal Systems, Phuket Marine Biological Center, 19–24 January 1998, Phuket, Thailand, p. 25.

Silliman University, 1996. Assessment of the Central Visayas Regional Project-I: Nearshore Fisheries Component. Final Draft, vol. 1, Dumaguete City, Philippines.

Society for Ecological Restoration (SER), 2002. SER International Science and Policy Working Group. The SER Primer on Ecological Restoration. (www.ser.org/content/ecological_restoration_primer.asp).

Soemodihardjo, S., Wiroatmodjo, P., Mulia, F., Harahap, M.K., 1996. Mangroves in Indonesia—a case study of Tembilahan, Sumatra. In: Fields, C. (Ed.), Restoration of Mangrove Ecosystems. International Society for Mangrove Ecosystems, Okinawa, Japan, pp. 97–110.

Spalding, M.D., 1997. The global distribution and status of mangrove ecosystems. Intercoast Network Newslett. 1, 20–21, Special Edition #1.

Spurgeon, J., 1999. The socio-economic costs and benefits of coastal habitat rehabilitation and creation. Mar. Pollut. Bull. 37 (8–12), 373–382.

Stephen, M.F., 1984. Mangrove restoration in Naples, Florida. In: Webb Jr., F.J. (Ed.), Proceedings of the 10th Annual Conference on Wetlands Restoration and Creation. Hillsborough Community College, Tampa, Florida, pp. 201–216.

Stevenson, N.J., Lewis, R.R., Burbridge, P.R., 1999. Disused shrimp ponds and mangrove rehabilitation. In: Streever, W.J. (Ed.), An International Perspective on Wetland Rehabilitation. Kluwer Academic Publishers, The Netherlands, pp. 277–297.

Streever, W.J., 2000. *Spartina alterniflora* marshes on dredged material: a critical review of the ongoing debate over success. Wetland Ecol. Manag. 8 (5), 295–316.

Sullivan, G., 2001. Chapter four. Establishing vegetation in restored and created coastal wetlands. In: Zedler, J.B. (Ed.), Handbook for Restoring Tidal Wetlands. CRC Press, Boca Raton, Florida, pp. 119–155.

Thibodeau, F.R., Nickerson, N.H., 1986. Differential oxidation of mangrove substrate by *Avicennia germinans* and *Rhizophora mangle*. Am. J. Bot. 73, 512–516.

Tunhikorn, S., Round, P.D., 1996. The status and conservation needs of migratory shorebirds in Thailand. In: Wells, D.R., Mundikur, T. (Eds.), Conservation of Migratory Waterbirds and Their Wetland Habitats in the East Asian-Australasian Flyway, Proceedings of an International Workshop, Kushiro, Japan, 28 November–3 December 1994. Wetlands International-Asia Pacific, Kuala Lumpur, Malaysia, Publication No. 116, pp. 119–132.

Turner, R.E., Lewis, R.R., 1997. Hydrologic restoration of coastal wetlands. Wetland Ecol. Manag. 4 (2), 65–72.

Twilley, R.R., Chen, R., 1998. A water budget and hydrology model of a basin mangrove forest in Rookery Bay, Florida. Mar. Freshwater Res. 49, 309–323.

Vivian-smith, G., 2001. Box 2.1. Reference data for use in restoring Tijuana Estuary. In: Zedler, J.B. (Ed.), Handbook for Restoring Tidal Wetlands. CRC Press, Boca Raton, Florida, pp. 59–63.

Watson, J.G., 1928. Mangrove Forests of the Malay Peninsula. Malayan Forest Records No. 6. Fraser and Neave Ltd., Singapore.

Weinstein, M.P., Teal, J.M., Balletto, J.H., Strait, K.A., 2001. Restoration principles emerging from one of the world's largest tidal marsh restoration projects. Wetland Ecol. Manag 9 (3), 387–407.

Whitten, A.J., Damanik, S.J., Anwar, J., Hisyam, N., 1987. The Ecology of Sumatra. Gadjah Mada University Press, Indonesia.

Wolanski, E., Mazda, Y., Ridd, P., 1992. Mangrove hydrodynamics. In: Robertson, A.I., Alongi, D.M. (Eds.), Tropical Mangrove Ecosystems. American Geophysical Union, Washington, DC, pp. 436–462.

Zedler, J.B. (Ed.), 2001. Handbook for Restoring Tidal Wetlands. CRC Press, Boca Raton, Florida.

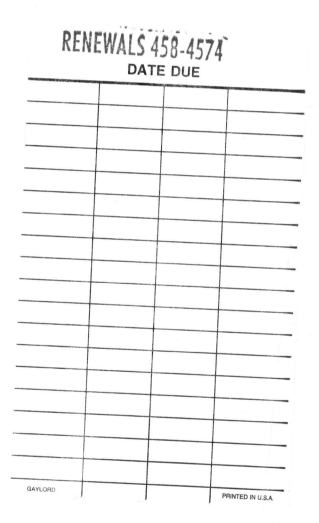